Stalking the Atomic City

Stalking the Atomic City

Life Among the Decadent and the Depraved of Chornobyl

Markiyan Kamysh

Translated by Hanna Leliv and Reilly Costigan-Humes

ASTRA HOUSE   NEW YORK

Originally published in the Ukranian language as *ОФОРМЛЯНДІЯ або ПРОГУЛЯНКА В ЗОНУ* (*A Stroll in the Zone*) © Nora-Druk, 2015. Published by arrangement with Agence Littéraire Astier-Pécher.

Translation © 2022 by Hanna Leliv and Reilly Costigan-Humes
Photographs © 2022 by Markiyan Kamysh

Photographs by Markiyan Kamysh, provided courtesy of the author.

Astra House
A Division of Astra Publishing House
astrahouse.com
Printed in the United States of America

Library of Congress Cataloging-in-Publication Data
Names: Kamysh, Markiyan, author. | Leliv, Hanna, translator. | Costigan-Humes, Reilly, translator.
Title: Stalking the atomic city : life among the decadent and the depraved of Chornobyl / Markiyan Kamysh ; translated by Hanna Leliv and Reilly Costigan-Humes.
Other titles: A Stroll in the Zone. English | Life among the decadent and the depraved of Chornobyl
Description: First edition. | New York : Astra House [2022] | "Originally published in the Ukranian language as A Stroll in the Zone, Nora-Druk Publishers ©2015." | Summary: "Stalking the Atomic City is a rare portrait of the dystopian reality that is Chornobyl. Focusing on the site as it is today, Markiyan Kamysh introduces us to the marginalized people who call the Exclusion Zone their home, providing a haunting account of what total autonomy could mean in our growingly fractured world."--Provided by publisher.
Identifiers: LCCN 2021050590 (print) | LCCN 2021050591 (ebook) | ISBN 9781662601279 (hardcover) | ISBN 9781662601286 (epub)
Subjects: LCSH: Chernobyl Nuclear Accident, Chornobyl', Ukraine, 1986. | Dark tourism--Ukraine--Chornobyl'. | Chornobyl' (Ukraine)--Description and travel. | Abandoned buildings--Ukraine--Chornobyl' Region. | Extinct cities--Ukraine--Chornobyl' Region.
Classification: LCC DK508.95.C545 K36 2022 (print) | LCC DK508.95.C545 (ebook) | DDC 947.7/709048--dc23/eng/20211214
LC record available at https://lccn.loc.gov/2021050590
LC ebook record available at https://lccn.loc.gov/2021050591

First edition
10 9 8 7 6 5 4 3 2 1

Design by Richard Oriolo
Map illustration by Jake Coolidge
Map source data © OpenStreetMap contributors, openstreetmap.org/copyright
The text is set in Utopia. The titles are set in Helvetica Neue.

To Flamingo

Contents

Stalking the Atomic City

1.

The Downpour

BRIMMING WITH THE OPTIMISM OF UTOPIAN slogans and the motherfucking grotesque of Soviet supergraphics, we were building a Dream. And in pursuit of it, we stumbled upon the Cornucopia—the energy of the peaceful atom, the panacea for the national economy, and the beacon guiding us on the path toward a bright red Communist tomorrow. Thrilled by our own might, with an undimmed belief in all that is best, we were building nuclear power stations all across the USSR.

One of the most powerful ones was called Chornobyl. Its satellite town soon grew rapidly, its neat apartment blocks towering in their exemplary excellence, enormous slogans flowing high, proud, on the rooftops, and boisterous children running around cozy playgrounds.

A supermarket and a restaurant opened in town, and ads like "Looking to exchange my apartment in Odessa for one in Prypyat" no longer surprised anyone. In the wilderness of the Polissya region, the Atomic City looked like something out of a sci-fi novel promising rapid growth, further improvements, and outrageous opportunities. They even planned to build a promenade with bridges, street lights, and musical entertainment. They already started to lay the foundation of new power plant units, the apotheosis of joy and happiness looming on the horizon.

Until things got fucked, and nuclear reactor No. 4 blew the hell up. The area by Chornobyl lit up like the Wormwood star and turned into a poisonous emerald in the precious crown of Polissya. The cruel hangover of reality after long years of sweet dreams. The law of the pit: no matter how long you climb, you'll fall back to the bottom in an instant.

However, brave firefighters put out the fire in the reactor, and valiant helicopter pilots showered the hellish crater with lead and boron. Desperate liquidators with pure hearts cleared the most polluted debris in the world, built the sarcophagus, and then left.

They left after they'd picked up their doses of radiation, their health problems, their cancer, their category A and B liquidator certificates, and so on down the list. Their children acquired the privilege to hang out at summer camps for free and to go by the nickname "Chornobylite" at school. The country got a piece of land as big as Luxembourg where people were forbidden to live.

The town of Prypyat and its surroundings were evacuated immediately. The Exclusion Zone was fenced off by barbed wire and patrolled by watchful soldiers. They raced around like predators on their armored vehicles in search of looters, but when the turbulent 1990s exploded with even greater force than the reactor, the Zone's borders loosened.

That's when the first illegals appeared. Haggard drunkards would steal pickled food from the cellars in the villages just outside the Zone and run away from the patrol guards only to come back in a week, get caught, and be thrown in jail—no probation. Prypyat was packed with daredevils, bums, deserters, looters, and fugitives. They hid in the villages for months, munching on rotten apples and dreaming of hunkering down until all the troubles of the world melted away. It was then that the Zone turned into that dangerous place often depicted in today's tabloids.

You could run into some hippies, too. Stories about flower children sporadically appeared in the newspapers—the police would catch them laughing and swimming in a river and kick them out with a stern warning: "Don't you ever come back, ever." Hooligans from the capital dropped by, too, to loot clocks from Prypyat apartments and peddle them at the flea market on Andriyivskyy Descent. They'd shoot up drugs and carry guns. Then the hooligans left. They left behind their meth trips like a whirlwind of ashes and became family men, completely ordinary: small business owners and loving parents of kids who are now littering your social media feeds with pictures of their breakfasts.

There were loners, too. They never left any footprints and drank good brandy. They fished in the rivers just to see the sun in the clear sky—they didn't give a damn that no one lived there and that they could be arrested. That's how it went until the generation the same age as the explosion grew up. To them, the Zone became a land of tranquility and frozen time.

I am one of that generation.

WHAT IS THE Chornobyl Zone today? For some people, it's a horrible memory of their half-forgotten childhood, of their happy Soviet youth, when, in a matter of days, their life shattered into pieces, and they and all their neighbors scattered, hopping on the evacuation buses to search for new homes. For others, the Chornobyl Zone is a pile of radioactive shit cleared away in May 1986. For some, it's a terra incognita full of myths about zombies and soldiers riding dark green armored vehicles. For others, it's authorized tours with greedy vendors delivering lofty speeches and making money on spaced-out tourists. For some, it's the backdrop of a popular computer game about macho men with Kalashnikov rifles who scarf down canned meat and bandage their gunshot wounds amid the fog of early-morning swamps. And still others believe that things are all bad there and see the Zone as a site from the movie *Chernobyl Diaries*.

In my case, it's even worse. For me, the Zone is a place to relax. Better than the seaside, the Carpathians, the gob piles (waste material removed after mining), or the Turkish resorts drowning in chilled mojitos. Countless times a year, I am an illegal tourist in the Chornobyl Zone, a stalker, a walker, a tracker, an idiot—you name it. They can't see me, but I am there. I exist. Almost like ionized radiation. What does it look like? I pack my backpack, catch a ride to the barbed-wire fence, and dissolve in the darkness of the Polissya woodlands, clearings, the pine-tree scent, vanishing in the dizzying thicket where no one can see me.

I'm talking about stalkers. Not the ones who collect children's gas masks in the district bomb shelters; not the ones who take pictures of unfinished piss-stained buildings in the residential areas. Not those stalkers. I mean the boys and girls who are not ashamed of shouldering their backpacks and treading through cold rain to abandoned towns and villages where you can guzzle down cheap vodka, smash windows with empty bottles, curse way too loudly, and do other things that distinguish living towns from dead ones. I mean the ones who are not afraid of radiation and don't turn their noses up at drinking water from poisonous streams and lakes. The ones who take awesome photos from the rooftops in Prypyat that later find their way into *National Geographic* and *Forbes*.

SOMETIMES I THINK that we don't exist. Not a single one of those forty people rambling time and again through Chornobyl's swamps. We used to exist, but we dissolved a long time ago in the mire and decomposed into duckweed, reeds, and sunlight. We are swamp ghosts.

Even flies do not notice us: they buzz around, busy, and pass us by. In the minds of our fellow citizens, we are a dim reflection of lies told on TV, just a bunch of tall tales about radiation, zombies, and three-headed calves. In the lethargic twilight, we spend hours looking for shallow places to wade through the impenetrable swamps; in the daytime, we drag ourselves along, up to our waists in leeches.

I just came back. Last week was a journey through the darkness, anxious anticipation of headlight beams and cigarette glow. There was hope for a bed, even without a mattress, and icy water from a frozen river; there was bitter cold and the quenching of thirst. There was the patrol guard I noticed at the last moment. There was grass: crisp, dry, and yellow. And then, only a deep sleep to run farther to the north in the morning and dive into fantasies, into a beautiful land of abandoned houses, canals, and agricultural machinery.

My fingers fumble on the keyboard, the cursor dashes madly across the screen, and the taste of the Snickers and Pepsi I scarfed down at the Polissya bus station doesn't align with reality. Only a gulp of a well-known brand lager brings me to my senses. Drags me by the lapels.

Not bad for 8:00 a.m. on a Thursday. But today I can do whatever I want. It happens to everyone. Archaeologists call it post-expedition syndrome; geologists have some other name for it, too. Every job that requires lengthy separation from civilization has a name for this condition. Yet I wouldn't call what I do a "job." I don't think much about this anymore. What I really want is two things: a bottle of good beer and a steak, well-done.

Where to begin? Perhaps with a good tip: never put ninety pounds' worth of stuff in your backpack. Don't go on thirty-day hikes or lug around enormous inflatable boats on top of forty cans of food. I keep repeating this mantra to myself over and over again, but I'm a masochist.

Sometimes you just have to be. When you hike in the Zone for three days, you keep thinking that it's almost time to head back. When you're there for five days, you regret spending only two of them in Prypyat. If you're there for a week, you think, "I wish I'd stayed for another day." No matter how many days you hike, you always recite the same mantra: "So, in eight hours I have to leave Prypyat and run toward the barbed wire. So, I have to be at the bus stop by five-thirty in the morning: a shuttle bus will pull up, no one will catch me red-handed, and the cops won't hassle me about coming out of the Chornobyl woods." You're always in a hurry. Always—except for the moment when you drop your backpack and walk around an abandoned village in the armpit of the Belarusian border, and you know you'll stay there for another three days. That's when you aren't in a hurry. This time, you won't rush. It will be a glorious trip. And a very long one at that.

There will be paved roads at night, the skeletons of power-line towers, and abandoned villages, fairy-tale-like in the nocturnal fog. During the day, a scorching-hot reverie reaches its heights there; at night, a damp mist settles on the ground and wraps everything in a web of gray shadows. Nothing disperses the mist, nothing disturbs it: only the light of headlamps, a streak of the Milky Way, and an infinite, star-filled glow.

I WILL REMEMBER these bright moments of my mad youth in my irascible old age . . . Here I am scrambling through the wreckage of chairs, through the cold insides of deserted houses, through lost life and trash that belongs to no one. Here I am conquering all the peaks of this stupid Zone: Prypyat's roofs, Chornobyl-2's antennas, the top of the reactor unit of the fifth energy bloc of the Chornobyl Nuclear Power Plant (NPP), the forestry tower, and the carcass of a looted bus turned upside down. Honestly, I don't know where to begin. Perhaps with the downpour.

I WILL NEVER forget that downpour. All of us have walked in heavy rain, so heavy that it's hard to breathe. All of us have once been caught in a downpour that topped our personal rating of rain showers on Earth.

At the time, I was drifting toward Prypyat in the dead of night. I was marching past the turn to Chornobyl NPP's fifth energy bloc. There's a crossroads there—a convenient place for guards to park their car and catch people like me. Stalkers, idlers, illegals, walkers, tourists—call them what you want. I pricked my ears and tiptoed as silently as I could until I ran smack into some officers—right there in front of me. I was lucky that it was 2:00 a.m. and the guys were sleeping inside their warm car. A straight road toward Prypyat lay ahead of me. And then, a downpour.

The downpour was hanging in the night air, with its thunderclouds like black ghosts, its skies blinking with dangerous electric flashes. The Chornobyl NPP was to the right, very close, some half a mile away. Lightning struck the horizon right behind the pipe of the fourth energy bloc. And then again. It was that old pipe that all illegals are now mourning because it was taken down—the symbol of our escapades that served as a backdrop for our time spent hanging out on Prypyat's rooftops was shamelessly dismantled.

What's the most important thing in the Zone? Right, the pipe. A phallic symbol of disaster. Smelly and sweaty, you arrive at the cold concrete high-rises in Prypyat; you break into an apartment, pick up the chairs lying around, drop down onto a couch, toss your backpack on the floor, and stare into space as your candle flickers orange. You make hot chocolate, go out onto the balcony, and light up a Camel, looking down onto the courtyard that turned into the impenetrable Amazon rainforest. Then you lie down for a nap, but in half an hour you wake up, cold and shivering, and you pull yourself together, crawl toward your sleeping bag, and cram yourself into it with your clothes on. You conk out and begin snoring loudly, testing the durability of this crumbly high-rise. You see the pipe in your dreams. And once you are all rested up and ready to go, you feel like getting drunk. You hit the bottle and chase it with some canned meat.

So, you trudge to another rooftop in Prypyat, pull a bottle out of your backpack, and start boozing. You lay out sandwiches with pâté on tar paper and pour a round, and slowly you and your fellow travelers—whom you dragged along for a whole day through shitty Chornobyl swamps—become the sincerest, the best of friends in the world. *Konnekting peepl* . . .

In these unforgettable moments, wistfulness, lyricism, and whatnot overwhelm you. You stare at the warmest rays of the yellow-hot disc in the far west, but then the sun goes down, and you return to that goddamn reactor pipe anyway. You remember the time when it flashed its lights against the black sky, a signpost for tired tourists amid the wretched overgrowth in the dead of night. Tourists plodded down deserted roads, and once they saw it, they knew: Prypyat was close, home was close. But then it was dismantled. The pipe was history. *Sweet dreams, dear pipe.*

But when I was waiting for the downpour, the pipe was still there. It was 2012: the year the world was to come to an end, the big soccer championship, and endless walks to the Zone. That night, lightning bolts struck every second. I charged ahead, and heavy raindrops pelted against my raincoat. I no longer dreamed of a dry evening in an apartment; I didn't dream of a nice little cup of tea, with two spoonfuls of sugar, hastily boiled on a camp stove before going to bed. I didn't dream of a dry sleeping bag. I dreamed of not drowning. But when I reached the border of Prypyat, I started to drown.

There I was, suffocating in the rain but slowly crawling forward against blasts of wind, under the high-voltage power lines. Every five seconds, a deafening thunderbolt crackled nearby, the blue veins of lightning cutting through the sky's blackness. Two minutes, and I was soaking wet—to hell with my German raincoat that was supposed to be water repellent. I felt as if I had just swum, fully clothed, across the Kyiv Sea; as if all the buckets of water from every ice-bucket challenge ever had been poured on me.

Prypyat loomed ahead—the destination of rookies who have yet to gain an appreciation for the deserted villages and are eager to take the litmus test, collect calling cards from all the abandoned places around the world. I'm not sure if I was a rookie back then, but it was my fifteenth trip to the Zone. There I was, ducking under a bridge to hide from the bright searchlight of a check-point. Old, well-trodden stairs took me down right to a railroad crossing, but I couldn't stand still during the August deluge. I ran across the tracks, trying to keep my balance, gasping for air between the raindrops, and I fell down on the huge, sharp rocks by the embankment, yelling the world's worst curses. I screamed with pain but the rain muffled all the sounds, and I couldn't even hear myself shouting.

There's a happy ending, though, even with all the cursing. I made it through the brush, soaked to the bone as

always, telling myself that I would never go back there ever again. I pulled a pack of cigarettes out of my pocket—by some miracle it was dry—and tried to light one up. While I fumbled for a lighter, the cigarette clenched between my teeth got so wet there was no way I could light it. I was on Ogniev Street, but there was no *ogon*, no fire. Up to my knees in water, I plodded toward the harbor—downtown, where an apartment and a cup of hot tea were waiting for me. Nothing was going my way. My sleeping bag was friggin' soaked. What kind of raincoat could possibly protect you from a downpour like that? But the good news was—I now had every right to warm up with some vodka. Good night, world . . .

Once I left all the downpours, canals, reeds, and fragrant swamp abysses behind; once I outran the Chuhaister, the spirit of the Ukrainian forest, and his legions of marshland demons in a marathon through the brush, I decided to spend the night amid the makeshift comfort of Prypyat. I pulled the duckweed, plastered all over me like emerald scales, off my coat and out of my pockets. Meanwhile, my beat-up sneakers were slowly drying by the cheap Chinese propane camp stove, which glowed quietly. Then, all the downpours in the world stopped, the star-studded sky threw its blankets over the creepy wasteland and the abandoned villages, and I fell into a sudden, infinite silence.

IN THE LIFE of illegal tourists, Prypyat is not the only desti-nation. There's also Chornobyl-2: gigantic pieces of iron, enormous radars, whose metal masts stick up five hundred feet closer to the sky. The ruins of a previous world. It's full of magic—moments captured by a spellbound, slumbering sun. The rust-and-star-colored steel legs of these unknown, sky-high giants make you wince, fall down on the ground, claw at the air, and dream of living next to something this huge. Forever.

In the spring, I always take magical naps by Chornobyl-2, basking in the rays of the March sun—cool, pale, yet already persistent. I find a fold-out cot in decent condition and pamper myself a little. I take a can of Pepsi out of a hiding spot and I put the cot on a roof and soak up the sunshine, sipping my cold beverage and daydreaming of a radar across the horizon, whose infinite masts leap toward the endless, bright blue sky.

After sunset, I wade into thick nicotine tar and fountains of liquor. Then I wake up at 3:00 a.m. Staring intently into the darkness, I try to get up and remember the previous night's calamities and the particulars of my alcoholic fervor. A dense blanket of cigarette smoke hangs in the room in the glimmering moonlight. My flashlight and smartphone have been dead for a while. When you're alone in the darkness that feels like a smoke grenade, all kinds of nonsense start to creep into your head. I run away from the smoke into the next room. Dragging the fold-out cot, I stumble over a dead wolf that's been rotting away under the peeled-off wallpaper for a year or two. I fall down right on its carcass. As it happens, it smells like

tiramisu. That's where I fall asleep. I love the Zone, mind you, I'm just not crazy about dead wolves. Sleeping right beside me.

IT WAS CHORNOBYL-2 again, and I was banging on a piano not far from the place where I had left my things. I can't play very well, but I placed the piano book above the keys nonetheless and opened it at random—"The Call of Winter." As if on cue.

So I requested "The Call of Winter" and it knocked me out like a bone-tired tourist. In the morning, I woke up to the sounds of someone playing the piano. My first thought was "It's her." But she was right there, lying next to me: hot, tan, and naked, breathing peacefully in the morning air. We shared the dirty hardwood floor, the cold sleeping bag, and the aroma of a life far from glamour and fake smiles. So it wasn't her. Who was it then?

"The Call of Winter" fell silent, and I grew anxious as I heard someone's footsteps next to our cozy nook. I got up quietly, took a seat on a stool, and lit up my first cigarette of the day, awaiting our guests. The Zone is such a funny place—someone played "The Call of Winter" so that I had to get up and sit on a stool, anxious, while she's sleeping—hot, tan, and naked. While I was waiting there for our inevitable encounter, I sliced the silence of the early-morning Chornobyl-2 with the creaks from the stool under my anxious ass.

But everything worked out in the end, and in a couple of hours, we went out again to search for secret swamps. Neither an inflatable boat nor a GPS or an extra pack of cough medicine could've helped us. It's just destiny. Someone freezes their butt off racing on skis, someone grows stiff selling things at a Chinese market when it's five below, and someone searches for fords. It's your destiny, your life, your desire. To float around in the abyss of fever and see colorful nightmares, greener than marshland duckweed, blacker than leeches, and hotter than my face when fever- ish chills knock me off my feet onto an old mattress in a village along the border, in the ruins of collective farms, among radioactive scrap in the thicket of predatory woods.

There was a boat, though; two, even. We brought the first one along with us, and we found the second one in the village of Horodchan in the middle of the night. We were searching for a house in good enough shape for us to climb up to the attic and drop dead asleep. The following day we were planning to go to Chapayivka and take a look at yet another village in the armpit of the border. At a bullet-ridden bus stop sign, at the gutted medical station with neat charts, at the destroyed school and the obligatory monument to the soldier-liberator. But this would hap- pen tomorrow. For today, we broke into a suspiciously tidy house with clean couches, and I happened to step on a floorboard that made an empty sound. Turning my headlamp up to the max, I pulled the board aside and saw an old rubber boat, a pump, and a pair of paddles. There was no dust in the hiding place, and the boat was old-school, which hinted at the shady dealings, of its owners. Why would tourists want to catch carp in this wilderness anyway?

This could be a simple fisherman—I'd say hello, bum a smoke, talk of this and that, and trudge on to damn Chapayivka—or it could be a grungy looter, or even three of them, and I doubt we'd part amicably, not with me half-awake, and her lying next to me, hot, tan, and naked.

You could scare them off, to be sure, but first you need to look like a madman, the kind of guy that people immediately point at when asked to pick out a serial killer in a group photo. Your appearance needs to frighten them so much that they stop cursing and give up the last Vatra cigarette in their pack. They would just pull all the cigs from their breast pockets and hand them over to you ceremoniously, turn off their headlamps and vanish into the darkness—vanish in the quest for copper cable, warm makeshift stoves, and the world's most colorful metals you couldn't miss even in the black Prypyat night. They'd run away to their nests somewhere out there. To the places where the rust of ancient swings nods at the murky sky; where the withered leaves on the basketball courts serve as a carpet and an orthopedic mattress. Somewhere in the jungle of the faraway villages where a soldier in his stone coat stands petrified as the birds clamor. The

Emperor, the Archmagician of the Swamp Brotherhood: calm in the May showers, unruffled in the August thunderstorms. Somewhere, the fallen power-line towers stick up like skeletons of long-forgotten monsters. Where looters built their nests from scraps of Soviet newspapers, black-and-white porn magazines, and packs of cheap Vatra cigarettes—gruesome, still lives of stolen happiness among forgotten ruins.

2.

Hot, Tan, and Naked Winter

WINTER IN THE ZONE IS A charming season, when the hearts of illegal tourists beat faster. There are magical moments woven from snow, wolves' howls, and soaked feet. Winter heats us up and makes us dizzy. We know how stupid our escapades are, but still—we take our backpacks and grab warm clothes and fifteen pounds of extra socks to stomp into an endless blizzard. To lose ourselves in huge snowdrifts.

With Her methodical destruction, Winter lets you clutch the fluidity of time. Time has no qualms about ripping the

tiles or scraping plaster off walls and showering it on land-
ings in hundreds of Prypyat's high-rises. But it is She who
works the hardest. It is Winter who punches the wall tiles
and scratches the faces of old Soviet murals. In the winter,
dampness descends, black, on the artifacts of the past: it
knocks over chairs, blows up hardwood floors, strips off
wallpaper, and tosses it onto the floor. She won't stop for
a second.

And that's when you notice the murals: they crumble
right in front of your eyes, tumbling down as small pieces
of colorful plaster. You come to one and the same place
time and again and see fewer and fewer artifacts of previous eras.

When I first found myself inside Chornobyl-2's school,
the mid-September sun was shining in, and shards of
broken glass crunched under my feet. The sound made
the birds flutter around the empty classrooms and perch
on the twigs of birches blossoming onto the windowsills.

The paint-chipped school hallways were decorated with
a mural of the soldier-liberator striking a dignified pose
against the red backdrop of blood and victory. Implacable,
time had already gobbled up his legs and was closing in on
his head. Next time I come back, I'll see only naked walls
and numb concrete where the mural used to be.

THE CHORNOBYL WILDERNESS in the winter is like Christmas during a blizzard. A long time ago, on Orthodox Christmas Eve, we plunged deep into the wild. Me and another tourist with an itch for adventure, a lot of baggage, and a raging wind in his hot head.

How did that trip start? Our friends came to see us off, and we sat on our backpacks in the middle of snow-covered Podil, a neighborhood in Kyiv. They tried to talk us out of going and scare us with the horrors of Winter, but booze emboldened us—two illegal rookies eager to go through a very strict rite of passage. But it almost went off the rails when two angry police officers with a muzzled dog caught us drinking alcohol in a public place. They let us go, though, after we insisted, nodding at our backpacks and saying that our train was leaving in an hour, and that if we missed it, we'd come to the police station anyway—where else would we spend the night in the middle of January? The officers left and we set off toward the Zone, though things felt ominous.

We had some tea at the Polissya bus station and then jumped into the shuttle bus for Ovruch—a shuttle bus going to the forsaken north, to dirty, small-town hotels where you stay once and never come back. We were supposed to jump out at Ivankiv but I fell asleep. Dead asleep like a blind hen, I fell down hundreds of miles into a soft seat, into an abyss of carefree dreams about abandoned towns and well-trodden paths covered with a thick layer of dry January snow. Into dreams about a happy Christmas. Thank goodness my buddy woke me up.

Ivankiv was a paranoiac's destination. Why, you might ask? Why not take the shuttle bus to the barbed-wire fence, say a loud "Thanks" to the driver, and get off under all the cursed, blinding lights? The drivers would rat us out. The moment we'd shut the door, they'd pull out their old Siemens phones and call the police captains they knew, turning in the wretched, unlucky tourists whose huge camo backpacks, with sleeping pads attached to them with Fastex buckles, would betray them from miles away. That's why we took a taxi.

We got off the bus and stood still, smoking in the darkness. My friend shoved packs of Camels into the side pockets of his backpack; I called a taxi. A moonlighter I knew showed up. "What the hell are you still looking for in the Zone?" he grilled us. I kept silent, but then I told him about the beauty of the gloomy houses gaping with the blankness of their broken windows in the middle of a numb Polissya night. I mumbled something or other about the landscapes of well-trodden trails and the comfort of staying the night in abandoned villages. About the cans of meat hidden in every corner. About the adrenaline rush everyone gets when police officers race around grabbing tired bodies—dragging them out of their sleeping bags and taking them to their squad cars, to administrative offense reports, to fitful bouts of sleep on the floor at the police station, and in the morning, a bottle of beer before getting on a shuttle bus for Kyiv. The taxi driver said nothing.

We all tell the drivers the same stories on the way to the barbed wire. The trip takes half an hour, and we just talk their ears off the whole time. Last week, this driver gave a ride to a couple of people just like us, and last month, he drove a Czech guy with a guide who took people to the Zone for money. Back in 1998, he looted hard until they slapped him with one year of probation for taking two tons of scrap from the Rossokha vehicle graveyard.

We arrived. My friend asked the driver to kill the light inside the car, and then he jumped out to check our surroundings. He came back ten minutes later and paid the driver, and he and I prepared to get out. Then the taxi driver turned on the light and blinded me. That was the last thing I saw before the darkness of the night forest and the depths of the January snow. Everything was all right; no one would be lying in wait there: only the nighttime asphalt, the blizzard, and the Milky Way on the black shirt of the sky.

We scrambled out of the taxi, trying to be as quiet as possible, since the checkpoint was right over there, some six hundred feet away, no more than that. "Merry Christmas," the taxi driver said. It was the seventh of January, two thousand eleven.

Wading through the snow, I was telling old legends about the remote emerald swamps, the thorny under-brush, and the virgin deserts of these lands. Somewhere nearby, a railroad bridge stuck out its dragon-like back, a towering ghost of distant epochs and heroic deeds. Brave firefighters in shiny helmets fought off the dragon's attack, and it fell down into the Prypyat and fizzled out in its cold water forever. Its skeleton was still hovering high above the water, scaring off foreign delegations and bus tours. People lay railroad tracks behind the dragon's back and forgot all about the hellish fires.

We forgot about the fires, too. We sank up to our ears in snow, cursing the whole world.

WHENEVER I WENT to the dump that is Chornobyl, I always dressed like an angry bum from Pol Pot's guerilla unit.

What did I wear that time? A pair of ripped-up Zara boots that I'd bought with some of my scholarship money many years ago, in my student days, in the times of historical documents, boring and fascinating theses, and lectures—some during which I dozed off and others during which I participated. One of the heels was just a gaping hole. I used to take those boots to get repaired three times a year, but then I gave up and just decided to put up with it. Bums should be colorful characters, so if your shoes have holes in them, they better be from Zara. I wrapped bags I'd bought at the supermarket ("Two large ones, please") around my feet and put my boots back on. I was hoping to keep my feet dry. So naive.

It always feels weird to climb over the barbed-wire fence in the winter: the snow crunches under your feet, the bags rustle in your boots, but you listen to those sounds and don't even care about being ambushed by the police. It's no wonder. After all, what kind of an idiot would stand in the middle of the Chornobyl forest when the snow is waist-high and it's twenty below? So, everything's all right, as long as you don't yell.

I squeezed carefully through a crack, clung to the barbed-wire fencing, and listened for my friend's quiet cursing, and then, all of a sudden, my phone started ringing, and I jerked to the side, got tangled up in the wire, and tore my jacket and my backpack. I was hanging on the fence in subzero temperatures, buffeted by cold winds, above the nameless ravines, in the blizzard of the Exclusion Zone. I was hanging like that for a long time. Thank God the booze hadn't worn off yet.

IT WAS SCARY back then: my first time in the Zone in the winter. My fifth trip. We were supposed to be chased by wolves, their paws touching the Chornobyl snow, the bottomless pits of their eyes gaping at us. Their huge eyes would've flashed if we shone our flashlights at them. We were supposed to light a flare, drop our backpacks, and step back to a tree. Climb it as high as possible, tremble with fear, cry for our mommies, call the checkpoint, warm up our hands over a lighter running low on gas, and freeze to death. The wolves were supposed to sit under the tree till morning and then, with the sounds of the first cars on the Kyiv-Chornobyl highway, go back to the shadows of the snow-white forest, growling and thirsting for revenge. This never happened, though. Of course it didn't—why would anyone care about us? Two drunken fools out on the road to Chornobyl in the middle of the night.

We made it to the asphalt road a while ago. It was always like that. You popped out somewhere between the checkpoint and the village, climbed over the barbed wire, ran across the meadow with all its pits, and stumbled, cursing, toward the ARSMS—a station that measures radiation (God forbid I'd have to write out its boring name in full here). Then the asphalt road. Six miles straight. We rushed headlong to our destination, pressed by time, the cold, and our fear of the wolves. Chornobyl-2 was up ahead.

As it turned out later, it wasn't just Chornobyl-2: there were also piles of snow, a pack of wild dogs on the premises of a factory, and unexpected cars cutting through the blizzard, startling us with their bright yellow headlights. We ran away from them onto the shoulder and fell into the embrace of the snow; we stuffed handfuls of it into our mouths to quench our dry winter thirst, to save water for the tea we were going to drink once we arrived at our destination. Next came a trek, hip-deep through banks of snow. Frightened, we began to sing songs about bears, half expecting to see wolves' jaws and fangs lined up behind every other bush, whiter than snow, whiter than all the January meadows of that winter.

In Zapillia, at twilight, we met a looter. We ran away from him, while he was running away from us. There was a restless night to the grunts of boars who'd left countless hoof prints around the house where we were sleeping like the dead. There were some real problems. As real as the taste of laundry soap, with seventy-two percent fatty acids. Seventy-two percent real like in between sore, wet legs, our sleeping bags, and vodka before bed. There was my friend's snoring and the anxiety in the morning, when we were supposed to set off toward our destination. And make it there. And we made it. But not before I fell down two hundred times and got stuck in the snow. With Zabara Lake ahead of us, we hid from roe deer, mistaking them for people, and pulled out the compass anxiously every other minute, scared of getting lost in the woods and roaming aimlessly until the night descended upon us and the sharp fangs of incredibly ravenous jackals, always lurking, sank into our winter jackets.

In happy anticipation, we sang again and again at the top of our lungs across the entire twenty-mile Exclusion Zone, and made a snowman on the road until we finally reached Chornobyl-2.

Steel giants, behemoths, titans—whatever you want to call them, those antennas five hundred feet high and two thousand six hundred feet wide in the middle of the remote Polissya woodland are the eighth wonder of the world, no less. The Petronas Towers? That's nothing. Imagine thirty Eiffel Towers in a dense row. Now this is the scale for you, and the scale of your wonder, too.

We threw our backpacks on the ground and started to climb in silence. At the height of a sixteen-story building, wind blew into our faces and our hands were frozen to the bone, as a frigid thaw stretched its wetness onto the antennas' rusty skeletons, and my gloves got soaked through. My friend captured the panorama of the expanse of snow on his old phone. You can't cram a hundred million impressions into two million pixels.

Under the vault of those incredible constructions, myriads of drops crashed against cold metal. Every moment brimmed with new sounds. You can't stuff myriads of falling drops into millions of bytes of voice messages. Even if Erik Satie played on all the pianos in Prypyat at the same time, he wouldn't have impressed me as much as those drops, wouldn't have beaten my hungover memory as hard with a sledgehammer of bright impressions.

We climbed down, burned an old chair, and made tea on it in the middle of the snow and eternal void. The trip back

home lay ahead of us, an endless night in the snowdrifts into which we were going to dive, eager and happy—we hadn't waded through the snow in vain. That time, I returned home absolutely content.

But all this happened a long time ago. A very long time ago. Nowadays, I walk to the Zone with a small backpack, bringing along three pieces of candy, two packs of instant noodles, a small bottle of vodka, a loaf of bread, a cup of lard, and a pack of cigarettes. I don't take anything else. Not

even a sleeping bag. It takes me exactly twenty minutes to pack up. I can race over to the Zone twice a week, and the trip by shuttle bus to the barbed-wire fence has become something like a run to the corner store for some beer. Nothing is more pleasant than the feeling of solitude in the nighttime forest where fear no longer resides. Abandoned houses make normal people sad, yet they make people like me sleepy and peaceful. I've found peace here. The peace of a country house I don't have.

IT MIGHT SEEM weird, but I always feel most comfortable spending the night in abandoned buildings. Be it a cold high-rise in Prypyat or an apartment without furniture in Chornobyl-2. I always feel calm in buildings where the roofing tar is covered with moss; where a ferocious May shower turns the roof into a swamp teeming with mosquitoes and tadpoles; where trees grow. Without rushing, I happily seek out high-rises in the jungle and find the ones I need: neat asphalt paths with curbs on either side, set straight as a ruler by urban amenities committees, leading up to the buildings. You can't walk there now—the benches are overgrown by pine trees, everything's hidden under a thick layer of moss and old leaves. These layers are thicker than the Earth's crust. Penetrating deep into the ground, the layers warm these abandoned places with the magma of memories, preventing them from freezing and from turning into an ice palace of emptiness and desolation.

I always relax when I roll out my sleeping bag among the clusters of stripped wallpaper. In the early spring, these clusters sag like shapeless heaps all the way down to the baseboards. I wrote my dearest wishes, curses, and dreams on that wallpaper; I put pieces of it, carefully torn, under the bottoms of not-yet-opened bottles, and when we boozed in the light of headlamps, in cigarette smoke and the aroma of canned meat, I knew it for a fact—no bad fate would ever befall me or my friends inside those abandoned walls.

Even in the rotten attics of Horodchan, an empty bottle's throw from the Republic of Belarus, the buildings bestowed peace upon me. I climbed up to the attics, unpacked my scanty possessions, and collapsed into a powerful, protracted slumber. I dreamed about Chornobyl NPP pipes, fires, police cars chasing us—laughing, we scattered all over only to meet a couple of hours later in a secret place we'd agreed upon earlier.

This was especially the case in Prypyat. It's huge and tiny at the same time. Tiny—for travelers and tour guides who built 3D maps in their minds of each nook over the course of hundreds of visits. They notice microscopic changes—last time, the plaster was still on the wall, and there was some scrap metal lying over there. Huge—for the checkpoint police officers who won't climb out of their warm dens unless they have good reason to do so. A guard won't roam around the city's neighborhoods with a flashlight, searching for looters. I doubt that he'd want to yell, "Stop, you fuckers, or I'll shoot," while holding a gun in his trembling hand. I don't think so. He's just like any other regular guy. That's why, at night, Prypyat is an endless array of bushes and petrified concrete, and these miserable hundred and fifty houses and thousands of apartments turn into honeycombs, into myriads of shelters for scrap hunters and nomadic bums who swarm here to scrounge the last bits of scrap, to booze, and to climb onto the roofs in the late afternoon.

The Dead City. Dead, oh yes. Twice dead. For the second time, with the emergence of thousands of photos and the shitty lines of official tours. Bored hipsters shot Prypyat dead with their expensive cameras; rich girls from the capital soiled the rotten couches with their tattooed backs and mapped every nook of the terra incognita on Instagram. The sense of mystery has been lost; it has escaped, vanished in the web. The aura of mysticism was scattered like ash in all directions, flying through the Internet byways to distant foreign lands. After that, abandoned apartments can't be scary.

ONE AUGUST, HOWEVER, I had a fitful night of sleep, at the barracks of a missile defense unit. It took me a long time to fall asleep. I had this vague feeling of anxiety, the kind that washes over you all of a sudden when flashlights switch off, candles burn out, and gas lighters go dark.

The place was unfamiliar, the room looked like a prison cell. And the door just wouldn't shut. So, there I was, lying in a Ukrainian-made sleeping bag in a comfortable temperature of forty degrees Fahrenheit. But I felt no comfort at all.

No furniture remained, and the most picturesque and cozy corner I could find was a room with steel bars—I wrapped a thin copper wire around them. It was empty and wet. Drops of water pattered from the ceiling onto the bloated linoleum, reminding us of the heavy rain that had poured on our vagabond heads the previous day, more than ten miles away.

I had a feeling someone would visit us that night. Not a tourist, not a wolf, not the police, but someone strange and dangerous. Someone who'd stir me awake with their resolute steps; someone who'd enter the room and turn on their flashlight without saying a word. The light would blind me, and what would happen next—only my fears know. In moments like that, it's hard to strike back: you're naked, and your sleeping bag is carefully zipped up. You're an embryo, a chrysalis, a miserable worm, and it would be a sin not to eat you up. But the most horrible thing is that she is lying next to you, and you're responsible for her.

I lay in silence, listening to the patter of raindrops. I played over paranoiac scenarios of all kinds and levels of horror, of all degrees of misery. I startled whenever a falling drop made a weird echo.

As time goes by, after your tenth walk into the Chornobyl Zone, you acquire odd habits and instincts. You can make out the different pitches of twigs snapping in the forest at night; you can distinguish the sound of winds in a pine forest from those in a mixed forest. You listen quietly to the sounds of drafts and doors slamming in the dead of night in the high-rises of abandoned towns; you fall asleep peacefully to the rattling and creaking of door hinges, since you know—dead houses tend to talk to their guests. They tend to bear old prayers and cries filled with pity in their souls. You know the sound of a raindrop falling on wallpaper, on linoleum, on rotten floorboards, on a hardwood floor, on whatever. You listen to these sounds over the course of long nights thousands of times. To the drops. You distinguish animals, too, how they walk—but the drops that night were weird, you know. They marched through the barracks toward me, and I jerked nervously in anticipation of a flashlight, someone's silhouette, and infinite darkness.

But no one came. Things happen. The gut feeling of an illegal tourist is a simple superstition you shouldn't pay attention to. Danger is real but fear is my choice, my late-night format. With these thoughts, I conked out, and I didn't have any dreams that night.

And then the morning came. Bright, all chirpy with birdies, warbly and joyful in the rays of the morning sun. The rays descended slowly on the plaster-spotted boards and the sleeping bag; they woke us up and reminded us that the Zone was not always gloomy and unkind. In moments like that, all your worries scatter and hide in damp, smelly basements—even if you had come there for the first time and had never left your house before.

We yawned, and then we brewed some green tea in teabags, smoked two cigarettes each, and destroyed a pound of canned meat. Time for a walk. Through the over-growth, through the hordes of ticks, through clouds of mosquitoes, through pine cones, old barrels, and cracked bricks. It is a special place. A two-hour walk from the anten-nas. But people rarely come here. Not surprising—who would be keen to explore the ruins of a typical anti-aircraft gun arsenal with empty hangars and scrap metal when there is the huge web of Chornobyl-2 looming right next to it?

The arsenal is smashed to pieces, old barracks blowing with cold air, and the hangars' skeletons glowing in the

morning light. There were lumps of iron everywhere, and on a metal pipe, long before the disaster, an unknown Soviet soldier, his hand steady, had written in red paint the eternal slogan: WE'LL BEAT YOU, REAGAN!

3.

A New Year's Eve Fairy Tale

WHEN YOU GO TO THE EXCLUSION Zone for the first time, you always feel like you should tag along with someone more experienced. And that's the first and biggest mistake. Bigger even than venturing into the Zone when it's fifteen below without a sleeping bag, hoping you'll be able to make a fire and sleep next to it, wake up at regular intervals, chop down all the fences within a three-hundred-foot radius to use them as firewood, and doze off again on a sleeping mat, cursing the whole world.

So, once you're ready for your first trip, go there alone. And most importantly, prepare as little as possible. You'll face true alienation: treading unfamiliar paths and sinking into swamps without a compass or a map, looking up at stars you know nothing about.

You have to be terrified of wild animals and take the shrieks of roe deer and moose for the roar of furious bears. You have to drop dead, collapse with exhaustion, but still crawl all the way to Prypyat, throw your bones on the ground and sleep like a log only to wake up, climb onto a roof, and realize—you've come here for something. You have to run around town with an illusory hope of finding a cigarette butt and taking a coveted drag, or searching for batteries to take a couple snapshots. You'll chase official tours and pick up cigarette butts like a real bum. After all, you're a bum, a resident of the apocalypse.

What do you need for your first trip? Nothing. Really— nothing at all. Take a flashlight, a knife, a few cans of food, a sleeping bag, a bag of rice, and a pack of lollipops. That's it. Of course, you can spend months researching the proper contents of a first aid kit and packing sleeping mats, flasks, multifunction tools, tents, and other tourist shit, but this keeps you from focusing on the most important thing. In abandoned towns, you have to rub your feet sore, drink tainted water, curse dead batteries, and save rice that seems to run out so quickly. It's much cooler, though, if you do this not because I told you to but just because you're a degenerate. Normal people have no business in a radio-active dump. Remember this.

YOU DON'T NEED any advice. Six months of packing, choosing a pair of pants to match the camo pattern of your backpack, a Flecktarn band for your Petzl flashlight—all this is merely an attempt to push your departure date back as far as possible. During that time, countless hurdles will crop up and disrupt your trip: you take out a loan, get stuck in a relationship, or finally start redecorating your house. Out of a thousand people, one hundred dig this idea and post about it on social media, a dozen pack their things, and only three actually go to the Zone. Go to Greece instead. There are too many of us anyway.

Don't forget that nobody's gotten rid of the radiation yet. My old man was a liquidator—dispatched for six weeks to the Chornobyl Nuclear Power Plant and Chornobyl at the time when you could still get fried by radiation. A civil engineer from the Nuclear Research Institute who worked with the neutron detection system of the Shater information and diagnostic complex. He went there on his own accord, not like the young soldiers.

He was horrified by their brazen fuck-it-all attitude—it was so easy for them to act all heroic and tear off their shirts when acute physical pain wasn't burning their asses. Radiation is like booze and drunken mischief. Got wasted, beat someone up, banged a girl. Sounds cool, but a hangover is worse than the Chornobyl disaster. It might turn out in the end that it was you who was beaten up, and that the girl got pregnant—and now you have a splitting headache to boot. Even a radiation crash course for dummies won't help the illegals—the absolute majority of them don't give a rat's ass about any norms or regulations, even as the entire bottom part of the periodic table sends an electrical charge through them.

Don't even try to say that you're special—lapping up unfiltered slime from a puddle with a sharp muddy aftertaste is far from extraordinary. After three days or so in the Zone, most of yesterday's radiophobes are eagerly sipping iffy infiltrate from the same little lake where they washed their feet after stomping on the liquidators' helmets and overalls that are still glowing, full of inspiration and vitality. Welcome to Chornobyl Land.

Some of us have long gone off the rails and flown into a cancerous abyss, and now we seek out the most contaminated sleeping spots, munch on sand from the Red Forest, and rummage in boxes full of radioactive junk packed with catastrophic background radiation, groping for fragments of graphite rods. Radiation fetishism serves as

a ceremonious rite of initiation into the caste of idiots—
the type of people who can get a Darwin Award while
they're still alive.

There's a radiology lab next to my house—you could
flick a cigarette butt out of my window and hit it. Just go
for it—draw a bucket of water from the Uzh, the Veresnya,
the Sakhan, the Ilya, from the basements of Chornobyl-2.
You've drunk hundreds of liters of it anyway. Over the
course of the past five years, over the course of sixty

trips, over the course of four thousand miles walked across the Zone, you've absorbed all the poison, all the background radiation and the radionuclides of Chornobyl Land, which has long become your home. Get that water tested and find out your level of fucked per liter of liquid.

PEOPLE OFTEN ASK me, "But aren't you worried?" It hasn't always been like this. At first, you spend three months researching, one month preparing, and another month worrying whether you'll get the chance to go to the Zone for a couple of days with someone experienced or not. When it's your fifth time, the barbed-wire fence barely even makes you tremble anymore. After your thirtieth time, you think nothing of dashing into the Zone for a short trip. You pack up in twenty minutes, and your backpack is disastrously small—a typical urban bag, not even a laptop could fit inside—and you pack stuff for three days. You don't take a sleeping bag. On scorching summer days, as soon as you get tired, you just fall down onto the ground and start snoring. There's no Zone, no radiation. No dosimeter—no radiation.

As time went on, we stopped shaking with fear next to the barbed wire and became friends with the local taxi drivers, looters, and border guards. The camouflage clownery and poker faces of first-time strikeballers fell into oblivion. Lightness, spontaneity, and ease came instead. We just packed up and took off for the weekend.

I'M PACKING MY things now. At 9:00 p.m. sharp on December 31st, another Zoneaholic and I will be standing in front of the barbed-wire fence. I have no doubts whatsoever about this trip. I strongly believe that no cops are going to be sitting in the naked winter bushes, shivering in their boots, on this festive, boozy night, waiting for tourists who are about to venture into the Zone to drink the liquor they'd stowed away and warm up frozen beer over a fire made in some hallway out of broken chairs and ancient wardrobe shelves.

I'm packing slowly. First, a case with a Pepsi logo on it. I put a can of Pepsi inside it and four candles (half a candle for each night—that should be enough for a week), a toothbrush, toothpaste, a pack of tissues, a needle, and strong, black nylon thread. With this needle and thread, I will embroider "fuck" on my backpack strap when it's torn to shreds at night in the middle of the winter forest.

Cards? Definitely. A pack of playing cards. I haven't brought a topographic map of the Zone for a long time—I should bring one some time. You're toast if you don't bring a pack of cards, though. Especially when you climb out of your den to check what's roaring on Lesya Ukrainka Street at 7:00 p.m., and the driver of a blue-and-white LAZ bus spots you because you're an idiot and you're wearing an orange sweater. No, you don't need a pack of cards just yet—what you need is to get the fuck out of there. The driver is tired and as angry as a dog. He'll curse at you, and in a minute, he'll arrive at the Prypyat checkpoint, and he'll tell a lieutenant he knows about the fuckers on Lesya Ukrainka Street. And again, you don't need the cards just yet. Again, you have to get the fuck out of there on angel's wings and run away from the neighborhood where you were spotted. The best idea is to duck into one of the highrises and run down to the mechanical floor. If you wait for just a second, you'll hear something like, "Woof-woof-woof-fuck, where are those fuckers? Kolia, can you hear them? Let's wait a bit." And that's when your pack of cards comes in handy. A pack of playing cards. So I always toss one into my backpack.

Then comes a gas bottle and a gas lighter: the holy duo, two saviors from cold, hunger, and boredom. When I'm about to conk out for the night, all I will be thinking about is my gas lighter. How I pull it out of its case, light a cigarette and warm myself up with some tea, drink booze, and stare at the burning banners made by red-hot flames while I throw another broken stool into the fire.

At moments like that, you don't think of clichés like "postapocalyptic" or "industrial." Actually, you never think about such stupid things. You just run around Prypyat for hours looking for a knife because you got drunk last night and fucking lost yours, and now you can't even open a can. Or you look for condoms, of all things. The funniest thing is that you do find them.

A sleeping mat takes up too much room. I did without one for a very long time. Just crashed on sawdust in the attics in the nearby villages: Novyi Myr, Denysovychi, Varovychi, Koshivtsi, Novosilky, Zymovyshchi. Just dozed off, collapsing on the time-chewed couches and snoring into the Polissya night.

A backpack is an important thing. A small one. You can crumple it in your fist, tape it up, and put it inside your main backpack. And then, when needed, you can toss in a pack of cigarettes, canned meat, instant noodles, a gas bottle and a gas lighter, a few Snickers, and a pack of playing cards. Basically, everything that will help you wait out the ambush as they hunt your stalker ass for several days. But none of us suffers from such assholism, and we do all this for the sake of one experience: hanging out on a rooftop, drinking beer, eating snacks, and smoking. We grab our backpacks, leave our warm sleeping bags on an early Prypyat morning and vanish among the naked bushes, frozen puddles, cold concrete, and gloomy high-rises. We vanish only to come back late at night, because we didn't just come here to booze and chill. We came here to . . . shit, why did we even come here? Well, someone came to take cool shots with their fancy camera; someone brought a group of tourists to earn money for their first car, a Daewoo Lanos. But why the hell did I come here? I don't like the industrial vibe. I don't like postapocalyptic stuff (it even bothers me when they sell unrefrigerated Coke at the supermarket). But I'll be damned if I don't come here a few times a month. Even if this month is a ball-freezing January, February, or D-ass-ember. Even when the snow is up to my balls and I'm wearing my old jeans. I'm here for some reason.

Even when the frigid snowdrifts of winter crash down onto the emerald-green swamps of Chornobyl, and the sun

vanishes behind the tops of pine trees without any senti-
mental farewells. When the twilight catches me unawares
and drags on forever. I embrace every spark of light with an
insane sense of appreciation. I trudge through the snow
across endless fields; I sink into the snow, pressing it down
to the ground lower than all the world's pre-dawn dreams
can. People take a breather and freeze to death in silence—
after the fifth drag, their knees start shaking and only
moving around can save them from the cold. I put out a blue
Camel in the snow, and at moments like these, I always ask
myself: "Why the hell did I come here and drag these people
with me?" I don't know.

THE FIRST OF January, four minutes into the New Year. I
pop the cork of the champagne bottle and launch a firework from a bridge across the Uzh. The last time I was
here was two and a half months ago—on my birthday. The
police caught me. They grilled me for a while, trying to
find out where the five people I was taking to the antennas were hiding.

Champagne gushes out like a geyser; half a bottle
splashes onto the asphalt, and I'm eager to have some boisterous fun. I'm itching to jump off the bridge into the cold

water, pummel the ice, and drown as the happiest person in the world. I'm just a loser who has nothing better to do than freeze his butt off among wolves and snowdrifts. Light snow is falling; it melts when you touch it, only a reminder of beautiful, enormous snowflakes. We finish the bottle as the firework crackles. The way you spend New Year's Eve is the same way you'll spend the rest of the year, as they say in Ukraine. I guess I'm screwed.

It's cold in Chornobyl-2, and someone boarded up my favorite hole in the fence, so I have to bash a new one open. Everything is how it usually is—silence and frozen ground on the flatland below the array of antennas, where grass never grows. The antennas disappear into the fog, their frost-covered metal bones dazzling the eye. We race along the typical routes of illegal tourists. I take pictures of fir trees—after all, it's New Year's. After all, it's the first of January, two thousand fourteen. After all, we're here.

I'm dreaming that right after one of these visits, they will saw down the antennas and destroy Prypyat with heavy machinery without warning the administration of the Zone and all those concerned. I really wish all this would pass into oblivion as soon as possible. So I could look at my old photos and remember those times with a sense of happiness. So I wouldn't regret anything in this world and would rejoice that I'd seen my fair share.

YOU THINK SOMETHING along these lines: "Well, let me just get to Prypyat, and to hell with the snow, to hell with the cold. I'll just get there, nestle up in my warm sleeping bag, and then roam around the town. I'll take the beaten path. I'll go by the Ferris wheel, pop into the Energetik Palace of Culture, and check out the supermarket and the hotel. I'll walk around the pier and take a couple dozen regular photos, and then I'll go to a comfortable apartment, drink a few pints of lager, fall asleep happy, spend a few days of my wretched life like that, and then march back home—to fix the problems I've created for myself."

The hell I will. I always think like that but every time, after yet another trip, I shake, hallucinating, wrapped in all the blankets I have, and I swear to God that I will never ever come to the damn Zone again, especially in the winter, because I'm a bum, a damned bum snooping around. I don't have proper thermals, a fleece jacket, or a soft shell, and I will never buy a sleeping bag that costs well over a hundred bucks—I don't have any of that stuff.

So instead of going for a nice walk around the abandoned town, enjoying the pastoral landscapes and the

spirit of the departed Soviet era, I head for Novoshepelychi. An unremarkable village, Novoshepelychi is my Mecca, my El Dorado, my you name it—there's a makeshift stove there. And only jackals go there, besides us. It's quiet there. And there's a stove there. A stove. A stove. I'm a complete and total bum, that's who I am.

A stove that I will keep burning for two days. I'll use up all the wooden fences in the neighborhood, and only then will I be happy. I will sit in the warm room and fart with the fire, smell with the smoke, gobble up canned pork, and wash it down with beer—and only then will I think, "It'd be nice to go wander around." After all, I trudged twenty-eight miles to get here. I'll crawl out of the smelly room, go out into the hallway, feel the temperature drop nearly twenty degrees Fahrenheit, and then I will say the hell with exploring all those landscapes, wrap myself up in my sleeping bag again, sit back down by the stove, warm up my chilled bones, and acknowledge the utter absurdity of these visits into this hole, into this bitter cold, into this slow, sullen death. I'm running a fever, and I stare, light-headed, at the ceiling, scrutinizing the patterns of cracks and old-school graffiti left by looters. Right now, all I want is to bounce back and run as far away from here as possible.

So we get the hell out of there. We just pack our things and abandon our previous plans to stay in the Kingdom of Fenrir for another three days. A big bottle of booze and four pounds of canned pork—we just drop them and go.

We absorb the last embers of the stove's warmth and then get the hell out of goddamn Prypyat. We walk down Lesya Ukrainka Street for a while; pass the checkpoint quietly, giving our enemies the middle finger in the dark; cross the bridge; and just run away, never to come back. We pass the Prypyat monument, a place where people on official tours love to take fancy photos, as car headlights pelt our backs; run away from that car into the Red Forest; and roll on the ground, feeling despicable. We reach the collective farm in Kopachi and make some hot tea inside a bus that has remained pretty well intact—well, at least for this shitty trash heap.

We make it out this time, and for two months afterward, I keep telling myself that that would be the last time. I am done with the Zone.

YOU THINK I grew wiser by the following year and opted for a noisy party in the capital? I wish . . . We went to fucking Prypyat again. Another New Year's behind the barbed wire with fireworks ten times bigger, and me a bit worried we'd be toast if someone took our festive fireworks for gunshots.

We had a thousand sparklers, ten holiday crackers, three pink ski masks, two Santa Claus hats, a sack of New Year's decorations, a fake fir tree, and a pile of glitzy garlands. And on top of that—four pounds of Russian salad, six pounds of Spanish tangerines, four LSD stamps, and the blessed confidence that we'd remember this New Year's for a long time.

And that was true. It was the most miserable holiday I'd had in a while. It was all about my karma. This time, I got served for all the years when I ran, apple-cheeked, ahead and spurred on by everyone else—faster, guys, quit dragging your feet! This time, the snow wasn't even waist-deep, but in some five or six miles, I was cooked, and my feet turned into one big blister, into clots of pain and nothingness. I felt as if I'd run a marathon in brand-new combat boots made in Ukraine, as if I'd treaded barefoot on all the asphalt roads in the world, as if I'd rubbed my heels with sandpaper, long and hard.

Later on, I wound up stepping on a rusty nail and started to ask to take a break every fifteen minutes. But I was not ashamed to do so, thank God. After five years in this shitty dump, there was no such thing as shame anymore. Whatsoever.

Finally, we came to Lubyanka, the oasis of the old Zone: the ghosts of dead grandmas still floated around there; clocks were in the cupboards, no longer ticking; jugs stood intact; and boots were lined up in the hallways. The village had not been overtaken by desolation and death, yet the houses were made up only of emptiness, fragments of chairs, and broken windows.

The paths of illegal tourists don't run through Lubyanka, and crowds of bloggers and photographers don't come around here, either. It still looks like the Zone of the nineties, those turbulent times when you went inside a house and you knew—someone had been living there just yesterday. They lived without water or electricity. They fetched buckets of water from poisonous wells, chopped wood—although they were already in their seventies—and always had a rifle at the ready, always had a walkie-talkie at hand to get in touch with the Dibrova checkpoint—in case something happened. In case those who left syringes in the nearby houses felt like taking something from the houses of the living. Gun control like they do it in America: no licenses, and if you trespass, they will open fire immediately and with great pleasure.

In one of her interviews, the Ukrainian poet Lina Kostenko said that she had a dacha in Lubyanka. A little house just for her—of course, she wouldn't tell anyone where it was. Until I found it. But then I went to Lubyanka again, cleared the table, fired up the stove, and realized that my dacha was there, too.

But, of course, I won't tell anyone where my dacha is. Neither will I tell you who I've taken there. TV hosts, actors, writers, famous artists—don't be surprised if you suddenly find out that your favorite singer or band pack their backpacks on the sly once a year and race toward the Zone after nightfall, searching for who knows what. Celebrities prefer

to cover up their Chornobyl hobby, and they don't post pictures from their latest illegal trip to Prypyat on social media, their dirty, cheerful faces against the backdrop of forsaken wonders. Tabloids or some other gossip dump would pay good money for a collection like that. Good thing I won't ever sell.

I DRINK VODKA at 6:00 a.m. Reserve your judgment—it's just the only liquid that doesn't freeze overnight. I walk around the house in my socks. My feet are so swollen with blisters that they won't fit into my slippers, and my sneakers are hot and wet, puffing out steam like the pipes of a power plant when it's five below.

We're waiting for Yura. I dropped some acid, and now I'm staring at the dragons inside the stove. They are calm and breathe hard, like guestworkers after thirty years of smoking red Magnas. One of them is ruby, the other one is emerald. I feel how closely they're watching me. I'm not too worried, but Yura doesn't know how to get here. He just takes off on the spur of the moment and never brings a map; he buys a few cans of gin, and once its pine-scented aftertaste kicks in, he's ready to march through all the swamps, all the peat moors, and all the impenetrable bogs of the world.

They hand me a phone, shouting, and ask me to tell Yura how to go around the Dibrova checkpoint. Should I even mention that I'm wholly absorbed by the emerald and ruby dragons? Hey buddy, there are two dragons at the checkpoint—be careful. I'm telling you as a friend.

Eventually, Yura will find us—he'll get to the bridge in three hours, and all this time we'll be waiting for him, warmed by moonshine mixtures and pleasant memories. We'll meet up, listen to the silence, slide down the hill sitting on a metal sign reading ILLIA RIVER, and go back to the house without saying a word.

I should climb up onto the stove and sleep there. The dragons won't find me for sure. They will fly out of the chimney and vanish through the windows. During the day, the stove got so hot that a thin sleeping mat from Bundesland, Germany, turned into melted cheese, and the sleeping bag shriveled up into something hard and useless. At night, the dragons flew out of the stove and tried to hunt me down, but they couldn't find me, so they flew toward Dibrova to watch over the checkpoint.

Lina Kostenko wrote that, to her, Lubyanka was a model of Polissya, a region that was fading away. A model, it's true. Sparse settlers drain oil from the last power convertor. Commendatore is still alive and insists doggedly that you tug on his balls—it cures every illness in the world, he says. Syringes still appear in the abandoned houses, and the locals still rat out random tourists. They just pull out their walkie-talkies and inform the Dibrova checkpoint about unwanted guests. And then dragons fly out.

This will all disappear soon. The last grandma will die, the remains of the oil will be poured out of the power convertor, and the crowds of tourists on official tours will snatch up items of the locals' everyday life and take them home as souvenirs. They will dismantle the mosaic of beautiful Chornobyl Land, and no one will ever put it back together again. In ten years, only fragments of photos, torn fabric of memories, will offer a glimpse of this outlandish world. And I know for a fact—I will miss it.

IT'S A PITY that we have to head toward Prypyat. To that doghouse with a makeshift stove—even a sergeant from the third squadron clutched his head in horror when he saw that shithole. It's about a twenty-mile walk there, and I can't help but think about my feet. I'm praying we stay put. But I pack my stuff and go. That stove is my only hope—it's just above freezing, and I don't have a sleeping bag or a pad. I do have fireworks, Christmas tree decorations, and a Santa hat. Happy New Year.

We'll reach the city in six hours. My feet sore, I will be standing in a dark hallway—without a flashlight or any hope for a nice evening. And at that moment, it won't be Prypyat, the Ferris wheel, the amusement park, or the Lazurnyi swimming pool that I will care about—I'll only care about when my buddies will come back from their hideout with beer and we'll move on to damn Novoshepelychi, to the stove and the mud. I won't stick my nose outside, and I won't give a damn about anything other than the orange-hot flames of the stove. Other than the dragons that save you from cold death, from the mythical wolf Fenrir's painful bites.

The stove hasn't changed—it's the same old filthy thing. Smoke gushes into the room; there's only a foul smell, trash, and a few rotten cots inside. There used to be a washstand, but I smashed it with a brick last year out of boredom and despair. We unpack, walk around the room, stumble over piles of trash, curse quietly, drive the smoke out of this damn doghouse, and eventually convene by the stove to thaw frozen beer and pass out. Then the wood in the stove will burn out, the night's cold dusk will move on, and the cold will settle with its lethal embrace around our shoulders—we will sleep tight and perish from the cold without waking up. The wind will howl with the lament of evacuees, snow will drift into the room, the whole town will be snowed under, and the snow will keep falling from the sky until the tallest high-rises in Prypyat drown in it.

There's no way around it. I pull Santa's hat all the way down over my ears, tell everyone I'll be back soon, pluck a bottle out of my backpack, and take a big gulp. Of course, I'm lying. I'll come back even angrier than I am now, and it won't happen anytime soon. I feel better at the bottom of alcohol delirium than in a reality made of dusk, snow, and smoke. Again, a less-than-glorious finale—the Mobius

rakes that I will step on till my dying day, knowing that, in the end, there's only the murk of Dytiatky, a nearby village, and a shuttle bus at 5:46 a.m.

We'll swear that *this* would be our last time. Next year, I'll switch off my phone, burn all bridges, drag a tree and a sack of tangerines home for New Year's, come across a miserable kitten, starving to death, and take it with me. The kitten will fall asleep under the radiator, and I'll play some ambient music and a slow Soviet movie and stuff my face with Russian salad. Do you believe me?

WE WENT COMPLETELY bonkers, burned down all the fences in our path, and gobbled down all our canned pork. I no longer believe that one day we will pack all of our things and get the hell out of here. This will be a moment of happiness. But right now, I'm lying down, sick as a dog, staring at the ceiling, and thinking about happiness—is it roaming somewhere in these poisoned forests?

The trajectories of happiness are more twisted than the sharp turns taken by kids playing tag on a hot summer day. Happiness was recycling bottles and getting a few hundred thousand karbovanets in return and then using that money to buy a lot of Fanta and Coca-Cola. Happiness was skipping school for a week to steal cherry plums from my neighbor's garden, listening to Detsl rapping on a cassette player, catching his new track on the M1 radio channel, and then recording it and hitting the replay button until my head started ringing. Happiness was mixing breadcrumbs with saliva, rolling it into a ball around a fish hook, and catching shiny carp in a small pit filled with water. Happiness was riding my bike over the hills, far away, and stealing the gifts of summer from enemy vegetable patches. Happiness was climbing onto my first roof, running through the subway tunnel from one station to the next, and flagging down a sleepy truck driver at 4:00 a.m. by the Dniester. Happiness was climbing rooftops and loving women, fighting on barricades and surviving, and getting a text message about a pregnancy test showing one line.

Fuck it. Real happiness is realizing that, one day, the barbed-wire fence will be there, and you'll stuff yourself with frozen canned pork, counting down the minutes till your shuttle bus arrives. Real happiness is getting out of there as fast as you can and forgetting all the horrors. Heading south, where it's nice and warm: blue Camels, yellow camels, suntanned Bedouins, azure oceans. Life.

I counted . . . and I got to around two hundred days. *Two Hundred Days in the Zone*—a nice name for a cheesy horror movie. Two hundred years, thirty-six thousand days, several million minutes among the cold concrete and hot swamps. All these high-rises buried underground a long time ago; they crashed and kissed magma, lingering in her fervent embrace. But I can't get out of here.

At first, we joked about our three-day stubble, then we just called ourselves Beard-os, but it's not funny anymore.

Now our beards are dragging on the floor, so you can just walk back and forth across the room instead of sweeping. Dust settles down in layers, and we draw short words on it. Like tiny letters in a cereal bowl lining up to form the word "death."

HOW WILL IT all end? Some of us will go explore Prypyat, while I will expropriate a sleeping bag and sink into eternity, without even a hint of dreams, without a chance of waking up by myself.

Some of us will make tea, pull out their cigarettes, and go outside onto the porch to have a smoke. A sergeant will happen to be driving by—a bike patrol accompanied by dogs. Well, hello. He might not notice anything and just whiz by, but there's a guy sitting there drinking coffee and smoking. Why not stop?

The sergeant will turn out to be a good guy. We'll drink gin and beer together as we wait for our shuttle bus; we'll get trashed, and he'll start telling stories about brazen Belarusian scrap hunters bumming tobacco at the checkpoint and how he kicks their asses.

Eventually, the police will arrive. All the officers will be in a great mood, a bit drunk and stoned; we'll set off some fireworks and palm off our New Year's decorations on them, uncork Kyiv-brand champagne, stuff ourselves into a red shuttle bus, and go the hell to Chornobyl. They'll write up reports and eventually take us to the bus station—to put us on the Chornobyl–Kyiv shuttle bus at the last moment, giving a strict order to the driver not to let us out until we get back to the city.

We'll drink hard, sitting there in our pink ski masks and singing ceremonial songs. The sergeant's visit was the most joyful part of our trip. I was already imagining how I'd limp twelve miles toward Chornobyl-2 only to drop dead, downright miserable, by another makeshift stove— my only chance not to freeze to death. So when I woke up and saw the men in blue honoring our bums' den with their presence, I rejoiced and uttered a warm "good morning" with a beatific smile on my face.

At the Dytiatky checkpoint, we went through the dosimetry control, jumping over the turnstiles and playing soccer with a bottle of vodka. Hello, mainland. We're going home.

4.

Campari

AT LEAST WE SAW PRYPYAT. IT could've been worse. For example, I could've taken some wankers from Barcelona to the Zone. I'm often asked what that is like—taking foreigners to the Zone? Easy.

They always have some funny names, like Samuel or Adrian, for instance. They drag me to Concord, and, once we're inside the restaurant, they grill me in broken Russian about the Maidan and Chornobyl and tell me how they went crazy in Donetsk during Euro 2012. They learned about the Zone from *Chernobyl Diaries* and they don't believe in monsters on the other side of the barbed wire, but they'd love to stare at the *dezertid chernobl ketestrofi* with their own eyes. They're just passing through Kyiv, of course, and listen, man, we got business in Warsaw, so let's head out tomorrow. Text us what stuff we should get.

Sania rolls up—their buddy, our interpreter, and just a dude interested in this dump called Chornobyl. I give him fair warning that, in the winter, the Zone looks scarier than a nude photo shoot of the minister of culture of the Luhansk People's Republic, but it doesn't help.

I slowly get more and more drunk at their expense, but I quickly sober up when I realize that I'm in trouble. One of them, right there at the table, flashes the gun they plan to bring along for self-defense. I spend half an hour trying to convince them that it's a foolish idea; for another half an hour, they try to prove to me that guns are fun, that we'll make snowmen, shoot their heads off, and laugh. In the next half an hour, they get as wasted as scrap metal, and I make Sania promise: no guns.

I meet them downtown on a cold morning. Sania arrives in his Toyota Land Cruiser. Thick snowflakes are pouring down from the somber sky. These macho guys have no idea what kind of arctic shit show they've signed up for.

They would never have signed up for this had they heard the story about Yura and the chocolate bar. That day, we all were psyching ourselves up. Someone was making a list of their achievements; someone else was trying to remember the faces of all their lovers; another guy was getting drunk on the shuttle bus so that his first contact with the snow would not throw him for a loop. And Yura got stoned and was craving a chocolate bar.

He tumbled out of the bus, plodded for some three hundred feet, waist-deep through the snow, and fell into a swamp—and then he just sat down, only his head sticking out. He refused to move unless he got his chocolate bar. We were swimming in snow for six miles to get to our overnight stop. It was freezing cold outside, and Yura, still starving, just couldn't pull his chocolate bar out of his backpack because the zippers had frozen shut. He had to make a fire to open his backpack.

We went inside the very first house we came across; he took off his soaked clothes, crawled into his sleeping bag, splashed some water onto the bottom of a bowl, groped about in his backpack for a lighter, and then turned around—and the water had already frozen. Yura didn't give a rat's ass about that, though; he fell asleep and woke up early and eager, well rested. He touched his clothes, but they were frozen; they had turned into colorful ice.

Samuel and Adrian were not worried. They came well prepared for all the snowdrifts of this wonderful January—they were wearing camouflage, proper autumn camo with anti-mosquito mesh. For most noobs, camo is a must. They choose a strict locally made oak-leaf pattern, desert Iraq, woodland, Flecktarn, or MultiCam. With their tactical backpacks, they meet at the Polissya bus station, but the local police patrol is on a first-name basis with the first squadron, and the bus driver calls the right person. His thumb strikes the old buttons on his phone, and he warns the patrol officers about the rookies—you could see them from a mile off. They blabber loudly about Prypyat, radiometers, and routes. Why not? They're wearing camo—no one will notice them.

We load our things into the car, and I flinch under the weight of these macho guys' backpacks. They took their preparations seriously and brought along eight gallons of Campari and five gallons of Chernihivske Strong. I know that mixing Campari with shitty local beer has an effect similar to the one produced by the explosion of the fourth power unit, and I gear up to watch it all unfold through the rose-tinted glasses of pure thrill.

We crawl down snow-covered roads and through Polissya's snowbanks toward the barbed wire, toward wonderful Trespassingland that will snatch these matadors and rub their noses into pain, snow, and despair. If you're in Kyiv and don't see any snow around, you'll get it knee-deep in the Zone. When it's knee-deep in Kyiv, among the overgrowth and meadows of Chornobyl Land you'll find snow up to your chest.

To the cheerful howling of the radio, we're approaching Karpylivka—the Mecca of rookies, and a trap set for them, too. It used to be a quiet place, until the Internet created an opportunity, a possibility, a chance. Until photo reports and route maps circled the globe. Until everyone learned how to get easily into the Zone through Karpylivka, without any trouble. The police learned about it, too, and started to snatch naïve first-timers in droves.

At the moment, I'm not worried. The snow is now knee-deep in Kyiv, so there won't be anyone freezing to death by the barbed-wire fence in this deserted void. Soon, we'll sink up to our waists in snow, and Samuel will finally understand that I was right and that no one could make it to the Chornobyl-2 antennas in two days—we'll just wander around villages only to return to our car half-dead, soaked, and exhausted, and we'll nod off while Sania drives us back to Kyiv. I warned them that it was a foolish idea to park their Land Cruiser there. I told them the approximate number of syringes in the abandoned house nearby, and Sania even translated the word "syringes" correctly. But he still left his car there because he wasn't afraid of the locals.

SAMUEL WILL SEE that the snow is truly waist-deep and that he no longer digs being a stalker. That walking eighteen miles to the antennas in two days sounds like a feat out of *Don Quixote*. He drinks half a bottle of Campari in one swig and passes the bottle to me, saying "pueblo." I'm all right with that, and I proceed to get sloshed. Then Sania walks—or, rather, swims—over to me, chest-deep in snow, and says, "It's okay, keep drinking," but what Samuel really meant by "pueblo" was some "village" that's not far from here. We'll look at it and then go back because it's a real mess here, buddy. Yes, I know you warned us.

All of a sudden, these two clowns start cawing something about *cierra puta*, get all riled up, and say that they've always dreamed of taking a look at the Rossokha vehicle graveyard: at its choppers with red stars on their sides, at its rusty armored vehicles and slender rows of LAZ buses.

All my warnings about the graveyard being like my lover—essentially nonexistent—fall on deaf ears, and they say, you're full of it, buddy, something should still be there, we saw the footage.

And when, three hours later, after three miles of snow and frigid horror, they see with their own eyes that the graveyard has turned into a completely open field without cartridge belts or aggressive Soviet military equipment, tanks and trucks, scattered about, Samuel starts guzzling Campari so briskly that I worry soon he won't be able to walk. Pueblo.

The worst part of it all is that the only way out, a miserable zero option, is the village of Rossokha. The worst thing you can see in the Zone. This village is notorious. Rookies who don't risk passing through villages and getting caught by the patrol officers end up here due to sheer frailty and despair after wandering through the labyrinthine structures of the Kovshylivka canals. Those bold enough to enter the Zone just outside Ordzhonikidze and go straight through the ravine that bears the sinister name "Boggy Moss" take this route here out of desperation. Once, it was ninety-five degrees Fahrenheit, and my backpack weighed seventy-five pounds, on top of that. I fell into a bog up to my neck twice. I plodded along waist-deep in duckweed and leeches, exhausting a week's supply of bug spray in just three hours.

Chest-deep in snow when it's eighteen below? Seven hours walking nonstop from Prypyat to the barbed wire? Sleeping in a plastic bag on naked concrete when it's twenty-four degrees Fahrenheit? That's nothing. Rossokha is forty-six times worse.

Even the floors in the houses are ugly. Old boards were ripped out to be used as construction materials, and you have to try hard to find a place where you can jump into your sleeping bag, zip up, and zonk out. The locals looted all the villages next to the barbed wire with the enthusiasm of the thugs from Toretsk who dragged fragments of the downed Malaysia Airlines Boeing to local scrap yards—like a carcass, a mammoth, prey, whatnot.

If most houses in Chornobyl Land look like an antique dealer's home blown up by a land mine, then Rossokha reminds you of a Potemkin village where only the facades are pleasing to the eye. Inside, you encounter a sullen, sober reality. A long time ago, I suggested tearing everything down and sprinkling the soil with salt so that people would never settle in this wretched place ever again. My Spanish buddies now have suffered trauma, and they have a horrible memory about the Zone—a hangover and desolation on par with that of Antarctica.

Of course, they don't like the village. And, of course, we turn back. They are tired and don't want anything else. Only Campari. With sympathy and regret, Adrian and Samuel grill me throughout the arduous, snow-covered trek back. Why did the graveyard disappear? And I keep hounding Sania, checking to make sure the EU citizens correctly understand the phrases "chopped the fuck up," "scrap metal," and "kick-ass loot."

And I feel so good thinking that we will all be sitting inside a warm Land Cruiser soon and this hellish, freezing-cold marching will come to an end that I guzzle down Campari at the speed of sound and the last thing I remember before passing out on the back seat is Samuel mumbling "pueblo" and sending a blood-red ribbon of Campari onto the sterile-white snow.

OUR SPANISH COMPANIONS continue to grumble about the graveyard that had been picked bare. They gulp down borscht with garlic rolls and drool over the never-ending rows of armored vehicles in decade-old photos from back when Rossokha was still okay. When the guards could fire a warning shot. When things were humming, and scrap metal streamed into the Potoky village scrapyard.

Only Buriakivka is left—a patch of land as big as a soccer field full of trucks, fragments of choppers, and the remnants of former glory. A meat-flavored substitute for life and a miserable parody of the things we lost.

But Buriakivka will soon share Rossokha's fate—scrap hunters have already arranged their gas tanks and torches to slice up all the metal and drag the heads of sad MAZ trucks back and forth across the graveyard. A few years, and that will be it. Goodbye, Buria, farewell, I will miss you, sweetie. I will recall, in a quiet whisper, all my drunken revelries in the skeleton buses amid torrential downpours. I will remember making love to all those girls who begged me to take them to this damn Zone. I won't forget lying on the roof of a bus and staring at the stars on one of those July nights when you can sleep comfortably in a T-shirt without waking up at 4:00 a.m. from a ball-freezing cold—a rarity for our temperate climate. I will remember all the dogs chasing me and the scratches from the rusted barbed-wire fences I squeezed through at the last moment as I raced toward the pine forest. I will remember all the somber shards of airplanes and moon rovers from the cleanup. They're sleeping. This is the last refuge of metal happiness; the last Klondike that will soon collapse under the pressure of demolition, under the pressure of scrap stockpiling plans and norms for the next five years.

It's a shame. One day, when I buy myself a mountain of cocaine and snowboard all the way down it, I will buy out Buriakivka, too, and will arrange all its vehicles around Prypyat. As decor.

But it's not Buriakivka or Rossokha that symbolize Chornobyl Land. The Zone has another symbol and it's perfect—the enormous masts of Chornobyl-2. Soon, savvy dealers will 3D-print tiny antennas, and people will hang them inside their cars instead of pine-tree air fresheners. Like little Eiffel Towers.

But the pipe was dismantled, so the Zone needs a new symbol. The worst thing is that the new sarcophagus may become the next symbol of the Zone—this huge, ugly garage, and I feel sick at the thought that, in a decade, a new generation will grow up considering this barbarity a symbol of the Chornobyl meltdown.

This will happen once all the villages fall into a state of complete destruction. When there are no more intact windows, when all the furniture has disappeared, when the illegal tourists have burned the last stool during yet another cold winter. Then you will no longer find a decent house to spend the night, and everyone will march toward the obliterated city of Prypyat and lounge around drinking tea on the rooftops against the backdrop of the new sarcophagus—an arch-looking pile of puke. It persistently reminds me of biodesign dating back to the turn of the century and conceptual art projects of would-be architects on the theme of eco-cities in the distant future. Building a thing like that in the Zone is even uglier than demolishing a historic downtown neighborhood and sticking glass skyscrapers in it.

BACK IN KYIV, as our Campari buzz wears off, our Spanish companions suggest we visit some looters. I am the one who pushes for the idea; I told them about that one time when my buddies and I were hurrying home and we bumped into a looter in the murky darkness of Prypyat at night. He was friendly. How could he have been unfriendly when he was alone and there were three of us? A guy with a headlamp and a walkie-talkie—on his way to cut up some metal without a permit. We smoked his cigarettes and told some yarns about peace and death, about metals, colorful and black, about fires and guards in hot pursuit, about life. And he told us how he was standing outside, stacking bathtubs one inside the other, and there was this fancy tourist from Amsterdam walking around nearby with her expensive camera. And how he got a kick out of his buddy, who threw another tub out the window that crashed down onto the asphalt below and shattered to pieces. Now there's someone in the Netherlands who will remember the Zone forever.

I doubt these macho types will be moved by the idea of a bathtub catharsis. There's a better chance it will be like last time, when a group of us walked to Prypyat and we saw those fuckers from Yaniv we've known for several years. They drive a crane from Ivankiv every week and tow tons of scrap metal. A man on a Verkhovyna scooter races around the city at night, hunting for remnants of colored metals. He has magnets in his pocket and his nickname is "America." Kolia America.

And we boozed in their kennel, gobbled boar meat, and got stoned on weed and drunk on alcohol we'd lugged for thirty miles along the railroad tracks from the glorious village of Radcha.

And then one of us boasted how much his knife cost, and the metal hunters exchanged looks. Then we all went to our Prypyat apartment and started boozing. Then we all returned to their kennel in Yaniv, and then, all of a sudden, the men in blue uniforms arrived. They packed us in the police car and took us back to the apartment to get our backpacks. But our backpacks were gone. Snatched. An illegal tourist, stoned as hell, shook the little police sergeant by the shoulders, howling, "They disappeeeeeared! They disappeeeeeared!" The little sergeant flipped the safety latch off his handgun. Everyone fell silent.

But we took revenge on those looters. Walking by, we let the gas out of their gas tanks, pissed in their hard hats, and stole their Ukraina bicycle. Then we got drunk and raced it around Prypyat until the front wheel went flat.

One day, we'll bring ten looters together, beat them up, and set everything on fire. Soon. But right now, just lie low. If you pass them when you're by yourself, just march on down Lesya Ukrainka Street and plop down on a couch, unpack your backpack, make tea, open a can, get drunk, and fall asleep. Crawl peacefully through the bushes, looking for a crash pad among the cloned highrises, among the fragments of window frames, broken glass, and moss on concrete. In ten minutes or so, you will find the gaping maw of an apartment building with a piece of the entrance canopy broken off—some drunk metal hunter threw a radiator off the top floor but missed. Now it acts as a landmark for you.

Everyone's asking me about it. In a world where firing grenades at Somali pirates has become a rich boys' game, everybody's interested in mysterious corners across the globe. In looting, too. People often treat me to an octopus salad, flying fish roe, and all kinds of delicacies so they can grill me about the degenerates from Polissya.

What do you need for a decent looting spree? A wheelbarrow is your best friend. You're in its friend zone. Listen to her whining about her exes and don't sneak a peek under her skirt. Just be her friend and bring her along wherever you go. You'll be able to take away much more. You'll just wander around dark Prypyat apartments without a flashlight, unscrew faucets, and stuff them into a big sack. You'll hoard a hundred and fifty pounds of stuff and throw everything into your wheelbarrow. Only its wheels will squeak when you start pushing it down the cracked asphalt. Don't roll it without oiling it first—the checkpoint patrol officers will hear the noise. They will chase you all the way to the Bridge of Death, and you'll reach go-kart racing speed, dump your wheelbarrow into the bushes, forget all your grievances, and hunker down at Kolia America's, boozing till dawn, smoking weed, and wolfing down boar meat, only to get stoned as fuck and make a triumphal entry into Prypyat on a scooter. Racing along its empty streets, sticking out your tongue, and drinking Chernihivske beer straight from the bottle. You're a bum. You're just a bum, and no matter what you say to these curious guys, you'll ruin these nice, curious boys that love their girlfriends and treat their friends to Beefeater on Fridays, these boys who just crave adventure. Don't ruin them. Just be quiet.

The Ukraina bicycle. She's also your girlfriend. But there's no friendship here—you two just have to bang regularly. You'd like to take away a sack of colorful metals and sell it to Alik the scrapper, right? You don't want to starve to death, do you? So you have to bang her long and hard. You'll carry your sack in the bicycle rack through the dead of night, and each sheaf of light will make you fall into a ditch, and every attempt to climb out will be as pleasant as hearing a machine gun burst just above your ear.

You'll lug your sack all night: all your dead cats, all your fears and worries will be inside. You'll lug it behind the barbed-wire fence and drink away all the cash you earned right there at the scrapyard. For a time, some local characters were paid in denatured alcohol, which causes a hangover similar to the Chornobyl disaster. When you get wasted on that horrible rotgut, catharsis will seize

you right on the spot, and your brain will be liberated by another reset. You're an old-school system, a Windows 98—every weird situation makes you reboot.

SO, YOU'VE DECIDED to go on a looting spree but you don't know where to start? Always pay attention to fragments of equipment. In the winter, bring a rope. When you get to Buriakivka, rip the hood off a truck. Then throw it on the snow upside down and stash everything inside it. Tie the rope to the sides of the hood and harness yourself to it as if to a plough—even a heavy hood full of scrap will slide well on the snow. Like a sled.

If there's no snow, you can stick some two hundred pounds of metal in two sacks into the rack of your Ukraina bicycle. Push it at night down an asphalt road or beaten dirt tracks—sandy firebreak paths are too hard to navigate. Always carry a magnet in your pocket—it will help you determine what kind of metal is in front of you—colored or black. Colorful metal is much more valuable. Even children know that. Make sure you don't miss it.

Don't forget about the elevators in Prypyat, either. An elevator shaft in an abandoned high-rise is not only a trash dump and a latrine, but also a springboard to financial enrichment. Of course, almost all the elevators were picked clean a long time ago, but do pay attention to the elevator coils. If you scrap a coil, you can go on a drinking binge.

Electric wiring and cables are no less important. You can find them in many houses, and they're colorful metal. By the way, make sure to burn off the insulator on the wire—then it becomes lighter and takes up less space. For this purpose, always have a bottle of solvent with you—you'll be able to make a fire faster, even if it's all wet, even if there's only snow all around you.

In the winter, spending a night without a sleeping bag when it's freezing cold outside can be a big problem. Only a few people know that you can make a fire right inside the room and sleep next to it. You need to break all the windows in the room first, and once you do that, don't worry: make a fire, and then go to sleep—the draft will disperse the smoke, and you won't catch a chill as you'll be lying right under the windowsill. The fire will slowly burn through the floorboards—they will glow for the whole night, keeping you warm. But they can smolder to ashes, and you might fall through the floor. But it's not scary—there's always a layer of sand under the wood flooring. In the winter, it's best to burn table legs, chairs, window frames, and fences—they're dry.

A TRIP IN the winter is a journey from woodstove to woodstove. Your routes and options shrink to a dot on the map, to the clusters of heat and warmed cells you cannot bear leaving.

But there are moments when even fires can't help. Then you start screaming. One time I started yelling, "Igor!

Igor!" Igor turned to look where I was, but I wasn't there. I had sunk shoulder-deep into the snow. Only my arm was sticking out. "Igor! Igor!" I was frightened. We thought that the snow was ankle-deep there. We thought it would be fun. In the Zone, in Chornobyl. It wasn't.

It took us an hour to walk some two thousand feet, and we no longer felt like being stalkers that day. Someone had offered me a pair of snowshoes, of course, but I flatly refused. It was either nothing, or a pair of Italian carbon fiber snowshoes, because "Italian carbon fiber snowshoes" sounded cooler than Karl Lagerfeld himself wearing Snowman skis—a Soviet brand—next to you. So I chose nothing. And nature chose minus eighteen degrees Fahrenheit. Natural selection wouldn't even look in our direction. We turned back. After swimming through the snow all the way back to the barbed-wire fence for an hour and a half, we went home. To hell with the Zone.

Bright moments happen, too. In the winter, you can always make tea, even if your flask is empty, even at the bus stop after you scramble out of the Zone, cold and hungry. You can thaw out several pots of snow and get dirty, ugly slush that even a stomach hardened by unfiltered mud puddles will reject. But it's possible. That day, I mixed that shit, took off my boots soaked through under the Niagara of sleet, climbed into my sleeping bag right at the bus stop, and slipped off into the dead dream of an exhausted tourist. I dreamed of the world's warmest blankets and came to my senses when I heard the shuttle bus rumbling. It always arrived on time, down to the minute. At 5:46 a.m. on the nose it sailed up to the bus stop by God's grace and redemption. The shuttle bus rumbled for five seconds and then drove the hell away. So I climbed out of my sleeping bag, gave up on my shoes, gave up on my socks, and charged toward the bus through the snowdrifts like a hardcore vagabond. At that moment, the police officers arrived, and they saw two bums running after a white bus, their sleeping bags clenched in their teeth. We made it. I didn't know what the driver said to the cops and why they didn't escort us off the bus and frisk us—I didn't have time to think about that, since I zonked out five seconds after I collapsed onto the soft seat inside the longed-for, warmed-up bus.

It's a teleport. Here you are—sullen, miserable, fucked-up by your night passage, crawling toward the exit. Here you are walking up to the piss-stained bus stop. You're swooning with joy. Your fears and misery sink down into the endless snowdrifts, into the crisp white snow trampled by wolves' paws. And when you start shaking with fever, when you fall all the way down to the cool Earth's core of

anxious dreams—the void comes. The chimeras that muddled about in your head leave you, and your thoughts fly away to bustling alleys and the Friday night rush of a big city. A single dot will come, too, and you'll stare at it. A single dot among the myriads of bright stars on the night sky. And this will be another reset. You'll return to

Kyiv, and people around you will move faster. You'll swim in the dimension of slowed rhythms. It will take you three days and nights to get used to the mad, big-city rhythm. You'll come home and fall asleep, an absolutely happy person. Nothing will disturb your dreams.

5.

Polissya Zen

I'M STANDING AT THE BUS STATION, in the corner, next to a trash can. This way, the patrol officers won't notice me and arrest me for smoking in a public place. It's drizzling. The drizzle turns into snow and then back into rain. It's 5:30 p.m., and she hasn't come yet. From a person going to the Chornobyl Zone with someone, I turn into a person going there alone.

This time, I brought along only a small backpack so no one points their fingers at me and says "tourist," "stalker," or "adrenaline junkie" through muffled coughs.

I'm wearing sneakers and a bright orange sweater so people think that I'm traveling to Ivankiv to visit my relatives.

Taking my time, I perform all the requisite rituals: I go to Silpo and buy a bottle of Coca-Cola and two of the chain's signature sandwiches. I look for the ones that have been sitting out for two days and are now discounted fifty percent. I'm not afraid of food. I drink water from utility tunnels, swamps, and puddles—I just can't get sick from stale sandwiches.

I buy a ticket. Route No. 4432, Kyiv–Hubyn. We're scheduled to arrive at 8:45 p.m., but we'll get there early. It will be dark inside the bus, and I will be the only passenger left by the final stop. The bus driver will play Russian pop music quietly. Once I had ticket number 5,000 on this very route.

The driver will keep silent, and darkness will fall on the tree line, the meadows, and the pale waters of winding rivers. I will gaze at the flames of the sunset and count the stars slowly coming through the soft violets of the night sky. I will stare at the solitary bewitching water towers and the skeletons of collective farms and villages whose names I will immediately forget. Stuck in notes of silence, I will await the perfect moment to break the law: to climb through the shreds of barbed wire and prick myself with the needles of pain and memories, the poison of looters' cursing and poachers' flapping. With the poison of life.

All this will happen, to be sure, but now it's time for the last ritual, the last sacrum—a visit to the bathroom at the bus station. If you show your ticket, they will let you in for free. What's more, you can go to the bathroom only once, and they leave a special mark on your ticket at the entrance—this ticket holder already took a dump. They'll draw a turd. A roll of toilet paper lies next to the ticket controller. And you mustn't tear off a big piece, or she'll curse you out. She doesn't let you take the roll inside the stall with you. At least she didn't let me. Perhaps she doesn't like me. After all, who likes a person who desperately needs to use the bathroom at a bus station?

It's strictly forbidden to fill up your own water bottle from the tap. They even printed that rule on a piece of paper. There's a bar of soap on a soap dish next to the faucet. There aren't any rules written about it. One time, I did fill up my bottle when I felt thirsty but had no money to buy bottled water. I filled up an empty Coca-Cola bottle while the ticket controller yelled at me.

It's a long walk to the water. I never took note of the exact time it takes. But it takes an hour and a half to get from Hubyn to Kamyanka if you're going at a leisurely pace. Then a thirty-minute rest at an abandoned bus stop. That's the Zone already. Then an hour of walking to the highway and another hour to Yampil Town Hall—a furnished building with a couch, a stove, a couple of rugs, and a guest book. Each time, I write in it, in Russian: "With thanks to Yampil Town Hall for its exemplary furnishings." I think it's a creative way of copying the style of inaugural speeches delivered at Communist Party conventions. Then it takes another twenty minutes to get to the water. You have to cross the Cherevach Bridge and keep your eyes peeled for the place where the black-and-white ledge ends. You can easily climb down behind it to the Uzh River. When you fill up your water bottle under the bridge and a car passes over it, you can see the yellow beams of its headlights through the cracks. The water is cold. I stir it with my flask and scoop up some sand along with the water. The flask has a wide neck. I'll throw it away one day. Plastic bottles are better.

I like the taste of the Uzh—bitter but without the metallic taste of the water by the pier in Prypyat. In the river under the railroad bridge, the water tastes a bit better than it does in Starik Creek, but it's not as clear as the water in the Veresnya River—the water in the Veresnya is just delicious. The brook in Kopachi is red, and if you make tea, better splash some Becherovka into it—it will smother the taste of rust.

Right now, I have only a can of Pepsi in a faux-leather case with a Pepsi logo on it. Like a nesting doll. Whenever I feel like cheering up the person I brought along to the Zone, I suddenly take out the Pepsi nesting doll and solemnly hand it to them. By that time, all the Snickers bars are gone, all the beer has been drunk, and you can only think about the twenty-five miles you have to cover overnight, the shuttle bus, and a bottle of Pepsi at the grocery store at the Polissya bus station. My nesting doll makes everyone happy.

There are lots of drug addicts in Zoryn. Today, they are dragging a cable out of the village—at 8:40 p.m., the power went out in Hornostaipil and Hubyn. Must have been the wind. At that moment, I was running through the nighttime forest. All kinds of worries, fears, and whimsies should have been visiting me, but I was only cursing the rain. Right under my nose, the wind knocked over a pine tree. I would've never have thought that you could die in the Chornobyl Zone as easily as that—not from beasts, not from scrappers, not even from a poacher's bullet, but from a fallen pine tree. But I didn't die. I didn't even get injured. I just muttered some curses.

Everything would be all right, eventually. I'd be sleeping the slumber of the dead in Kamyanka, the wind swaying the trees, rain flying inside through the windows, and the drug addicts from Zoryn lugging cable to Hornostaipil. No, not to scrap it, but to turn the power back on.

Now, I'm sitting at the bus stop painted yellow and blue, and an unshaven guy in sandals comes up to an old lady and tells her that the power is back on. It's 12:20 p.m. This has never happened here before, she says, power outages used to be rare. It has something to do with the storm warning. I know it, because I was nearly blown off my feet by a gust of wind. My backpack was soaked through, I was freezing, and I broke into a house I haven't been to for three years. We used to drink beer there. The house stood close to the KAMYANKA signpost. Since then, the roof collapsed, and someone dragged chairs and mattresses into the only room still intact. Three metal mugs were on the table. A typical looters' den.

The old lady says there used to be no Zone here, the local boys didn't shoot up, and lots of people would take buses out here on weekends to pick mushrooms. But then the meltdown happened, and now there are soldiers everywhere from Hornostaipil to Strakholissya. Even in Hubyn. She tells me about the barbed wire, too. During

the first year, the fence was equipped with electricity and an alarm system, and the brass stayed at her house. She complains that her son was arrested for foraging chanterelle mushrooms, slapped with a thousand-hryvnia fine, and taken to court. They never take me to court—I am a big-city guy.

She gets jealous and asks me about my parents. She's surprised to see young people come to Hubyn, to these old ruins—they don't shoot up, they are neat, they don't use foul language, and they buy houses there. Most of what she's saying has nothing to do with me. But she wishes her own son would be like that. They wouldn't have taken him to court then. They haven't taken me to court, have they? I don't pick chanterelle mushrooms.

The shuttle bus slowly starts off toward Kyiv. In Strakholissya, someone wrote PUTIN IS A DOG in black paint on a fence. At the bus stop next to it, a mutt takes a leak onto a profiteer's checkered bag—a man put it out on the curb so the driver would see that there's someone at the stop and won't drive past. The man put it on the curb and fell asleep. We drove past.

There are seventy people living in Hubyn. In 1970, there were many more, and a school was open up until 1975. Later on, local children took the bus to Hornostaipil. My ex called Hornostaipil (Weaseltown) "Shynshylivka" (Chinchillaville). I found that really funny, and I walked around repeating this word over and over like a line from some stupid song.

It's very quiet in Hubyn. A bearded man named Misha covers his motorbike with a tent. He grumbles about the weather, saying that it's going to rain again tomorrow and cold March raindrops are going to drench all his junk. No one listens to him, except for Vanya, the fool idling around by the Energetik Palace of Culture, bumming money from passersby. No one gives him anything. The locals know him too well, and newcomers hardly ever drop by.

Tania is fixing her roof. She's the only one with money for metal shingles. Or for new locks for her barn. An old woman scurries around. She grew anxious when she saw someone pouring vodka into plastic cups. No one listens to her at all. She dashes from house to house and asks for booze, she calls beer "champagne," and if you ask her to keep her distance, she showers you with all kinds of curses typical for Polissya witches.

Liuda is sitting in her yard. No one bothers her; she doesn't bother anyone, either. Her grandchildren are making a racket, though—their parents dump them off here for the weekend three times a month. The youngest boy has already explored all the mushroom places in the local woods. The grandchildren want soda but the grocery store is closed for a never-ending lunch break. The sales clerk shuffles over from the other side of the street—it takes her an eternity—strikes up conversations with the passersby, and tells them about the guys who came into her store yesterday with their large tourist backpacks to get bottled water, beer, and cigarettes. Must have gone boozing at the Kyiv Reservoir. Really, where else could they go? Then she goes into the store, turns on the radio, and eats the sauerkraut she didn't sell, and her mood brightens only in the late afternoon when a cold spring sun falls down to the edge of the sky. Another day has passed. Another day on the Chornobyl Zone frontier. I'll come back in three days, when they aren't predicting any wind or falling pine trees. They're predicting seventy degrees Fahrenheit, clear skies, and a friendly face in the pine forest. I'll be back soon.

I'LL BE BACK to visit Krasno. It has some charming nooks and a truly alienating vibe. It's utter desolation; no official tours race along its old asphalt roads. They don't build any sarcophaguses in Krasno or take pretentious pictures against the backdrop of mighty Soviet military equipment. The Zone is different here, away from the eyes of random tourists and hangouts at popular Prypyat apartments. And there's only this abandoned church with an owl and bees living in it.

I call the owl Armavir. After our fifth meeting, he's no longer afraid of me. When I remove an enormous chain from the entrance to the church and step inside, it sits still on the iconostasis, and if I were to come too close, it would take off silently, fly up without making a sound, and slip out of the gate. If I hunker down, it'll return an hour later, quiet as a mouse the whole time.

The owl scatters the donations, rips Belarusian rubles and Ukrainian hryvnia with its sharp claws, bites at the Communion bread and cherry pies, catches mice, and listens to the bees humming in the church dome amid the silence of numb, scorching-hot summer days. This buzzing echoes in the corners of the church, more soothing than all the ambient music on the planet.

Among the abandoned villages and forgotten houses of the frontier, among the swamps and the faraway corners of the Zone, there's an orthodox church with a clean floor, prayer books, and a donation box. With candles, Armavir, and drowsy bees. When I stayed there overnight, I dreamed about my distant childhood, about the sun and my carefree life—the most peaceful dreams in the world.

This is the place of my earliest awakenings and my most delicious coffees, the most pastoral sunrays in the whirls of dust and the sincerest good mornings. Soft and kind, their echo murmured up to the dome, stirring the bees and scaring Armavir away.

In the late morning, I removed the chain from the gate, and sunrays flew inside. The sun bestowed happiness upon me, and in the middle of this forsaken land, amid depression, syringes, and hundreds of layers of dust, amid the looted hearts of all things abandoned, this church in Krasno stood like an oasis of joy and happiness, an island of optimism and good thoughts.

One day, my cat died. I buried him deep, wiped away my snotty tears, tossed my shovel aside, and went to the Zone. I walked for a good forty miles until I finally got to Krasno. I spent those fifteen hours thinking about my cat. I remembered how I picked him up off the street, how he ran after me, meowing, how I brought him home, how he always walked around outside as much as he wanted, how he returned home with his ears scratched, and how a motorcycle ran over his paw and I borrowed five hundred bucks for treatment at a time when I myself had nothing to eat.

On my way to that church, with all my cats in my backpack, I was thinking that people light a candle in church only once—when someone dies. Our memories of our loved ones who are no longer with us fly away with its smoke—up to the church dome, to the bees and higher, through the cracks into the sky. The smoke will fly up to the sky; tears will drop onto the hardwood floor. The candle will burn out, and I'll forget everything. Farewell.

This church is dear to me, and I believe that out of all the things in the Zone, it's the only one with a future, with its patch of hope, sentimentality, and infinitely dazzling sun-rays rushing in every morning when I open the heavy gate and light fills a semi-dark vestibule.

THE ILLEGAL TOURISTS are not concerned about their own future or whatever might happen to the Zone tomor-row. We're obsessed with the moment that's fading away forever—year in and year out, we've been watching our dear Chornobyl Land crumble, the valley of our peace melt under the snow. In a hundred years, reenactors will play out the last day before the meltdown, charming Polissya's final moments. They will wear drab clothes made by the Donbas Textile Factory, and, in a rebuilt Prypyat, they will throw a festival on a scale as large as *The Truman Show* for each and every person and replicate the lives of past eras. Mass tourism into the past—the exclusive tour of the future. Welcome.

Actually, all these rosy dreams of mine are just an antip-odean alternative to the gloom of the present day, to the dusk of the fall. Day three here, and it seems that the murk has deliberately spread its sullen skirts and pressed down

on you with its cold clouds and drizzle. It wraps cotton balls of a morning fog around your throat and goads you with the whips of thirst through the bushes toward downtown. It goads you through the dew and wasteland toward the well-trodden stairs of the pier. I'll fill up my plastic bottles there and guzzle this metal-flavored liquid without worrying about the harm it can do. It's pouring outside. I fidget on the couch, listening to music through one ear pod, pull out my sleeping bag, and crawl as deep inside it as possible. The battery dies faster than the sun sets, hidden by the clouds and wrapped in a hundred layers of fall mist. I'm sitting like this with the balcony doors open, listening to raindrops pattering, wild boars padding around, listening to the rain, silence, distant metal banging, and the screams of illegal tourists. To the Vopli Vidopliasova of boys and girls with big, hot hearts dancing to the rhythm of big-city life.

I fall asleep sooner than the sun rises on the other side of the city, sooner than the cold glow of the sunrise falls onto the gray asphalt on yet another gloomy morning. I close my eyes, tired of being bored, and snore, dreaming about the bright future of these lonely apartment blocks.

When I wake up, it's still drizzling outside. There's a day ahead of me that promises nothing. The calm and peace that my soul needs are here; I hate thinking about going back. I'd better go look for some old books in the classrooms of the local schools. Otherwise, I'll fly off the handle and stay here to write insane monologues and magic spells in forgotten languages with charcoal on the walls.

I stumble across a copy of *Pionerskaya Pravda*, something by Gorky, and a tenth-grade physics textbook. I read them and then use them to make a fire in a back alley. I get bored stiff, and when the sun goes down and it becomes hard to make out the letters, I venture into the drizzle toward the gigantic metal cranes in the harbor and I catch the first drops of rain. The cold drizzle of fall. I've been hiding in the crane's cabin for two hours. Then l remember that it gives off two roentgens, so I climb down and trudge back. The rain won't let up, and I am soaked by the time I return. I make a fire out of old chairs and window frames—to warm up my joints and dry my clothes. Then I race around the neighborhood, scouring apartment blocks for furniture I can burn, to listen to it crackling under the awning, and so I can finally get ready to leave.

Orange flames perform a fierce dance, glinting off the peeling walls, and I open a can of pork with a knife, mash it up, and heat it until it boils, dreaming about a shot of vodka. Wretched rain. This gloom settles here every year; mist covers deserted autumn parks even in the capital, the streetlights flashing like lighthouses with their yellow glow. Only these warm patches of big-city amenities can help you crawl out of these labyrinths of gloom, fall into the softness of your own bed, and sleep straight through until the first hot rays of sun. But in the Zone, they burned out thirty years ago.

ON A GLOOMY fall day like this, in the glorious city of Prypyat, I met two looters. They were from Belarus; they pronounced "straight" like "stayt" and "fuck" like "fock;" and I couldn't help laughing hearing this baby talk from these two burly guys.

At first, they wanted to smother me, and I tensed up. I didn't lose face, though. We struck up a conversation, and they suggested getting a drink. Pretty typical story. At a small-town bar, a group of bodies come up to you and offer you a drink on them. As you're getting hammered together, some sort of problem arises, and they decide to knock your teeth in. I knew for a fact—somewhere under their soaked camos, Makarov handguns with scratched-off serial numbers were tucked away, heavy, together with ghosts from my eerie nightmares, the cold heels of Prypyat's drizzle, and the horrors of inescapable tortures. I said yes to Lukashenko's vodka and the Rahachow canned meat, to quiet tales told in the glow of headlamps in a dirty hallway, as we sat on rickety chairs I hadn't burned yet.

It was a conversation typical of people who forget their email passwords, who don't follow political news, and if they have to choose between going to the movies or to the sauna, always chose the latter. They complained about their dented UAZ—last time, the police almost caught them. The cops chased them and cut in front of them like in the movies.

The meatheads told tales about thermal viewers and the Belarusian border guards. And then they shook my hand and went away to start their UAZ, knock back a small bottle of vodka behind the wheel, smoking out their rolled-down window and racing as fast as they could northward, across the border, to exemplary cities without commercial ads. I would've left, too, but I came down with a fever after all those days spent in the cold twilight drizzle.

I often visited Prypyat when it drizzled and the raindrops pattered on the old window sills; when the apartment blocks drowned in the mist. But time and again, I saw the city in tender sunshine: amid an ocean of molten asphalt, amid the summer fervor, overgrown with apricots and apple trees, amid a jungle and attacking legions of insects. When dawn was full of metal clattering, dogs barking, and scrap hunters yelling. When there was a scorching midsummer sun, silence, and bugs buzzing in my ear. When I ran onto the hot afternoon concrete to count the official tourists in the square, to count the vapor trails among cirrus clouds. When I was sleeping.

6.

Hello, Zone! Farewell Forever!

SO I RETURN FROM THE ZONE and think to myself: "Here I am, back home, drinking my orange Hike, devouring brand-name chocolate bars, washing them down with Pepsi, and enjoying the spice of life." And then I realize that I have a few words to say to the folks criticizing consumer goods of modern civilization. To everyone who hates that you can buy ten kinds of frozen veggies and twenty kinds of cigarettes at the supermarket. You're fuckers. Just taste a chocolate bar after a two-week trip to the Chornobyl dump; just feel the hazelnuts crunching on your long-unbrushed teeth; just take a swig of soda and only then can you curse having access to a large variety of foreign goods. Fuckers.

So I'm sitting at home, watching Kubrick, drinking a little bit of tea, coffee, and Hike, cooking soup with garlic and mushrooms, eating it slowly and with relish, and then sleeping under all the blankets in the world, believing wholeheartedly that I never ever will go to that shitty dump, to that piss-filled Zone again.

When I was six, someone gave me several encyclopedias about nature as a present. They had a lot to say about tectonic activity, limestone sinkholes, and coral reefs—I read them over and over, hundreds of times. Later, when I was thirteen, I played through Max Payne thirty times—it was the only game on my hard drive. I didn't need anything else. Why would I? I had Max Payne 1 and I was obsessed with it.

Same goes for the Zone. For a normal person, one time would be enough, I guess. For a fanatic, a few times would do. But I found a new destination each time I went there. At first, Prypyat and Chornobyl-2, then the villages, villages, villages, pioneer camps, recreation camps, antiaircraft installations, hangars, railroad tracks, cooling towers, and churches. I wanted to sniff and touch every patch of this dump, every fragment of the past. And every time I came back, I swore that it would be my last, my very last visit.

I'd be damned if it were. In a month, here I was again: ripping my backpack on the barbed wire, stumbling into potholes in the fields, marching along the forsaken railroad tracks, crossing bridges, and lighting candles in abandoned churches. I'm an idiot. Please kill me.

THERE WERE MOMENTS of optimism, too. One sunny morning, I woke up in good spirits. Nothing was troubling me. I opened the windows and drank some green tea, staring at the city. Then I went to the gym to pump iron, spent twenty minutes in the shower listening to the sleep-inducing tracks of Brian Eno, got out, and ran to the supermarket. Pushing the cart across the floor, I bought a bottle of Pepsi, some cheese, ciabatta bread, a pack of hot dogs, a few cans of pork, and lots of German beer. I grabbed a plastic raincoat and ordered a taxi. Eventually, I crawled as far as the Klyvyny station in the middle of the Chornobyl thicket and zonked out in the open air, seeing sullen dreams until the morning lifted its blankets, until the cold dawn dragged me out of sweet oblivion and my warm sleeping bag.

During trips like that there are many warm, pleasant nights when all of us gather around a fire and talk for a long, long time until everyone falls asleep. Until we burn our sneakers because we've come too close to the fire. We chug liquor diluted by swamp broth, look at the stars, and stare at the fire, and it seems as if there's nothing more beautiful out there than the Zone.

The trip's storyline is as simple as shit: walk slowly, chase the whimsical thoughts in your head, and cherish hopes that soon you'll jump under all the blankets in the world and fall into a slumber for a couple of days. At home. In the dark, amid sweet dreams. Break through the bushes, stumble through the artery of an abandoned railroad that turned into a forest, into a land of wolves a long time ago. Its ties became beaches for snakes, while the paths alongside it are now tracks for wild boars. Fall down onto the gravel, take a nap in the rain, in the starlight, in the floodlight of the full moon. Crawl to a crossroads at dawn and nod off, only to wake up at home, in your shower, to the sounds of your favorite ambient. Only to jump up and take off in a couple of weeks.

MY ONE THOUSAND, one hundred and forty-sixth time. A good enough number to give up this shit. My feet are wet, my arms wobbly, the morning infinitely cold. The bus stop is empty except for the two dirty bodies wearing backpacks and smelling like medieval pilgrims. They smoke nervously, staring in the direction of the lumber mill. At 5:45 a.m., they'll hop on their shuttle bus—they have every legal right to do so—nod off in its soft seats, and fall into fitful, innocent dreams about the police, about being chased and ambushed unexpectedly. They'll fall into dreams about the smoke from a makeshift stove that brought all the Zone police to them. Dreams where the officers of the Security Service of Ukraine interrogate them only because they can't find anything better to do than roam through the overgrowth of the Chornobyl Zone. They'll arrive at the deserted Polissya bus station where, in the early hours of the morning, you can see only bums and women selling dairy products; they'll go down, stinking, into the subway, keeping an eye out for the patrol officers who could stop them, since their IDs have long expired, and the police officers could hassle them about so many things, not just about looking like utter bums.

Am I going to quit making illegal visits to the Chornobyl Zone? Do I want to lie in bed all day with a 102.2-degree fever after another absurd trip, after another bum voyage, dragging myself to the kitchen, barely able to peel an orange or lift a bottle of sparkling water? I don't know.

OF COURSE I know. Give me a month, tops. I pull out a laminated map of the Zone, twice the size of a regular sheet of paper. I had it with me when I ventured to the most distant corners of Chornobyl: amid backwater swamps, duckweed, emptiness, and mosquitoes. I spread out the map and started whacking off to it.

I just drink my morning tea, munch on some cookies, and turn all my attention to the topography, to the green patches of woodland, to the white fragments of meadows that have long run wild and turned into a dense maze of young fir trees, to the underbrush and swamps that I will forge through with the fiercest curses in the world on my lips.

I know that I will go there again, as there are blind spots still left on the Zone's map. I've been planning to go to the village of Vilshanka, but I haven't been there yet. I keep walking past it every time.

My friends come over often, and we spew non-sense about our shared past and funny memories, about geopolitics and literature. They drink green tea or down vodka—it doesn't matter. Sometimes, they stare at my map and, for the hundredth time, they look through the photos from the Zone that, for the hundredth time, I "treat" them to—they nod, deciding to go there with me. I never

say no to my friends. I know too well how it goes. Here they are sitting and nodding in agreement: "We should go, oh yes, we really should." They think that it's not for nothing that these trips are so popular and that it's not for nothing that *National Geographic* writes about fools like me. But none of them will go there. They'll find eight hundred reasons why they can't: my leg hurts, my girlfriend won't let me, I don't have time for a weeklong trip and I don't really want to take vacation time for something like that.

It's all right, I say, of course, it's all right—that's very responsible of you, you have to think it through and weigh the pros and cons, you have to decide if you're ready to expose yourself and your loved ones to such a risk. I understand you, of course I do. But you know what—fuck off, I've got my own plans, and I can't be bothered to take someone to Prypyat for the fuckteenth time.

AS A MATTER of fact, I don't like taking new people along. Or some of the old ones, either. I have to take them where we agreed to go. To Prypyat, that is. But what if on the way to Prypyat, somewhere after Chornobyl-2, it occurs to me that I haven't been to the Emerald summer camp for a hundred years and that its little cottages will soon crumble—what then? In short, when you bring someone along for a trip, it's like you've thrown a manhole cover from the ninth floor onto your own head—a manhole cover of obligations and rules. It presses down on you, and you don't feel the lightness that should always accompany you during your ventures into the Chornobyl Alienation Zone.

Or let's say you all made it to damn Prypyat. You've laid down on some couches to rest, drunk yourselves silly on Becherovka and tea, and walked all around the central squares and popular tourist destinations. Now you feel like moving on, as you still have a few days ahead of you. Of course, you could take your fellow travelers to the railroad bridge or to Krany, show them Novoshepelychi or some other place, but, personally, you'd love to race over to the church in Krasno, on the left bank of the Prypyat River, to light a candle, make some tea, and say hi to Armavir. But according to the laws of inertia reigning in the Zone, these lazybones don't feel like going anywhere. They think only about wretched Prypyat and won't consider anything else.

What's more, they just run out of gas. They run out of gas on mile twelve and, unwilling to admit that they can't hack it, they start telling some tall tales about sudden injuries and insist that we all walk at a leisurely pace. All of them hurt their left knee. All of them.

LET'S MAKE A deal. No "expeditions," "marches," or any other lofty words from the arsenal of homebred warmongers. "A walk in the Zone"—that's the only name that'll do. There's nothing in the Zone that could turn it into an ultra-dangerous place, into a strength test for the most resilient humans. Would you prefer a test like that? Then go to a Siberian swamp or climb down the mouths of volcanoes, because in the Zone, all you'll get is unhurried walks through mixed forests.

Humans always place obstacles in their way. Here they are whipping up hysteria, claiming that Chornobyl Land is exotically dangerous and dangerously exotic. Here they are watching TV programs about wild animals and looters. All of that to work themselves up, to be afraid of their own shadows, to succumb to imaginary fears and the secrets of illegal prohibitions.

But once you have traveled the lengths and breadths of Chornobyl Land, once the Prypyat skyline has turned into a mundane sight, an unremarkable backdrop for a late-night tea party, once you have climbed the Chornobyl-2 antennas so many times that you have lost track, once all the abandoned collective farms, villages, hamlets, and forestry towers have long been explored—then you start searching for things unattainable.

And you find places of utter psychosis with hot swamps where vagabonds sink and insects buzz. You no longer catch a shuttle bus to the barbed wire—you just hop in a taxi and climb into the Zone silently in the dead of night. You walk through the bushes that only you know, following the labyrinthine structure of the canals without a map, guided by your nose and sixth sense. You choose the most distant villages and corners of the Zone. But you end up going to Vilcha, anyway.

THERE ARE ABSURD and cruel things in life. You can deliberately refuse to have a drink the morning after to cure your hangover, you can race halfway across the city on public transportation every single morning, you can guzzle gallons of low-alcohol drinks, smoke hookah, pop the collar of your polo shirt, listen to Russian chanson on the radio, get married at eighteen, and take out a loan to get the latest iPhone.

But there's nothing worse than finding yourself in Vilcha. Well, the beginning of your trip is nice, even joyful. You tumble out of the huge bus that dashes toward Ovruch daily and then glide through the village of Radcha.

Radcha is a special place. A place of sleepy old women and screams from the bar where smugglers, looters, and border guards all booze together. If someone really wants to sober up, they'd better go fishing in Prypyat for three days. It's impossible to sober up in Radcha.

Then you'll find yourself in Vilcha, at the station branded by warm hues of Soviet supergraphics. A guy nicknamed Beaver lives there. He always has a ton of booze. Compared to the unshaven scrap hunters, he looks like he is the master of this corner forsaken by God and humans, and your best companion for a few nights full of sullen drinking. Beaver cuts up the remnants of scrap here, occasionally drinks with the watchmen at the cell tower, and draws water from a well that only he knows how to get to.

No one bothers Beaver. Only people like me drop by for a few shots of vodka. Beaver tells them about a hermit who lives nearby. He makes rafts. Beaver pours water from the well into old bottles and lets them go.

This is the Zone already: it's been the damned Zone for three kilometers, thirty years, and thousands of tons of metal. The Republic of Vilcha is a place of mutual understanding and respect. The police pay their respects to the old women—the self-settlers—and when they die and lie for weeks in their crumbling houses, the police pay their respects to the scrap hunters and promise them a reward—a three-day looting spree in the village if they agree to bury the corpses. Everyone respects Beaver. Beaver doesn't think about respect—he just boozes and gathers scrap metal.

I HAVE A special relationship with the police, to be sure. Here I am, tired, approaching a house in Prypyat after a long walk around the city. Here I am grabbing a can of pork before going to bed and wrapping myself up in the world's warmest sleeping bag.

The light in the window gives my buddies and me away. The cops wake us up; they stir our sleeping bodies, frisk us, and promise to kick our asses if we move or try to toss anything aside. I complain that they ruined the moment. The police sergeants fiercely insist that I show them my documents; they frisk me with tripled enthusiasm but find nothing. And as I'm still sleepy, they sneak a Swiss army knife along with my passport out of my pocket and yell at me to pack up my stuff. We stand at the checkpoint for a long time waiting for something; I smoke cigarette after cigarette, listen to the police officers scolding me, and stare at patches of the Milky Way, searching for familiar stars: for spots and signposts of joy, combinations of happiness and misery that can postpone arrests and all the reports in the world. Why would I feel upset? Fishing in the estuary of the Prypyat is next on my plan, anyway. And it's even easier to get there from the barbed-wire fence.

The Zone is different in the estuary. Shiny speedboats and enormous fish, abandoned floating factories and dried-up corpses of ships with funny names. Thousands of islands drown in the early-morning mist among the last breaths of an Indian summer. The sun throws its generous gleams onto the reeds and the dense swamp grass, and there are people hiding on these nameless patches of land. They hide there to wait out the patrol officers, to swim toward Prypyat, to catch huge catfish in the forbidden waters and then sell them to Kyiv restaurants whose names have sunk into the swamp.

Eventually, the police officers shove me inside their cruiser and lock me up in a cage where you sit bent like a question mark, atoning for the sins of wetting your whistle in public places or petty crimes committed in a big city. They also toss two backpacks into the back of the car, along with a pot of boiled buckwheat that we didn't have the chance to finish due to the police raid. The door slams shut, and the police cruiser starts up and rolls away from the secret sites and the abandoned cities. And we bounce up and down on Chornobyl's pothole-ridden roads.

I try to imagine some nice things, but I can't think of anything other than a bottle of alcohol. I remember that we have buckwheat and I try to eat it—it feels as if someone is giving me a tattoo while we're racing off-road at forty miles per hour. I grab handfuls of buckwheat, but it all falls onto my pants. I try to light a cigarette. We smoke a lot, in fact, but the last cig is always reserved for some absurd ceremony at the bus stop after all the running in the dark. The last blue Camel always hints at the upcoming trip home. Even in the police cruiser. Smoking the last Camel feels as good as throwing a bathtub into the fresh morning air from the fifteenth floor onto the asphalt.

Eventually, they take me to the local police precinct. The platoon commander mumbles something about spies and traitors. He calls for a colonel of the Security Service of Ukraine. The colonel, wearing hunter's camouflage, swims into the room and says in aggressive Russian that we are dead, but then he lets us sleep at the precinct. I'll get my police report tomorrow.

As a matter of fact, the fishermen from Chernihiv were supposed to bail us out—they are friends with the Chornobyl cops. Misha the Mackerel should've gotten us off, but he was boozing with a police sergeant from the second platoon, and then they sailed away. Maybe they crashed into a wave and sank. Misha rose to the surface. The cop did not. Well, Misha, you're a devil—drowning a cop.

Eventually, I go to the Ivankiv District Court. The sun is shining, and it's clear that the asphalt on the Chornobyl–Kyiv road badly needs to be patched up. I'm dreaming of a beer. In two hours, my sullen bum friends and I crawl out of the Ivankiv court, and the cops soften. They let us buy Chernihivske at a nearby watering hole and offer to give us a lift for three hundred hryvnia to our next drinking spree by the Kyiv Sea.

We all cheer up and jump into the police car, smoking Camels. The sergeant at the wheel turns up the radio. We huddle in the back seat. The police chevrons are dangling in front of our eyes, and a Russian chanson is blaring from the speakers as our ride flies through the village of Sukachi. It's all empty, only a body of what looks like a heavy drinker is lugging a huge metal pipe on his ancient Ukraina bicycle. He fuckin' snatched it.

HAVING AN ENCOUNTER with the police is much more pleasant than bumping into local wildlife. We stopped fearing wild boars and wolves a while ago, and bears are sporadic guests in our lands, but stumbling upon a lynx in the middle of nowhere is more frightening than all the police officers in the world, all the reports and interrogations by the Security Service. A megahertz of lynx. To the sound of thousands of heartbeats and predatory breathing, you build barricades, stacking doors, while lynx cubs play noisily in the attic and their huge mother lynx breathes heavily, the megahertz of her breathing on kindling Communist pioneer fires of fear and panic.

A long time ago, a cat scratched my face and broke the skin. Ever since, I've been having nightmares about tigers biting me to death. Once a month, one apparition always played out, again and again—it made me jump out of bed and stare, frightened, at the young moon through my German-made, dual-pane window. The moon hid behind the clouds. I tried to fall asleep but fell into another frightening dream. Here I was walking around Prypyat, and I bumped into a den of lynx cubs. My blood ran cold, and I screamed silently to myself, "Ru-u-u-u-un!" and, at that moment, something grabbed me by the scruff of my neck.

When the lynx came into contact with me in the Zone, she always retreated, but that time, she did not. Wheezing and rustling, growls and silent threats could mean only one thing. We were unlucky on two counts—we'd chosen a house where lynx cubs lived, the place where my nightmares lived. Terrified, we banged a stick against a cooking pot in an attempt to chase away the mother lynx who was forcing her way through the cracks in the barricades. We built a fire out of furniture in the middle of the room and started to suffocate, but dying from smoke fumes seemed much more pleasant than dying from a lynx attack.

The coals were hissing, the bodies of chairs crackling, and our fears, prayers, and pleas for salvation were curling up to the sky through the holes in the ceiling like wisps of Polissya incense. All my black thoughts and tribulations rose up with the smoke into the bright, starlit sky. When the morning tossed its grayness and melancholy onto the misty meadows, onto the damp mixed forests, when the birds' chirping announced the freshness of dawn, dew on the grass, and yet another warm, sunny day—the lynx fell silent. We fell silent, too. Into a dead dream.

That time, everything worked out for the best—the lynx that had been lying in wait for us in front of the house ran away the moment we went outside. Perhaps her cubs were sitting in the basement or in the attic. I know only one thing for a fact—death was following our footsteps that day.

One hardly ever wishes an unwanted guest in a blue uniform would drop by. But that day, I was beyond disappointed that it wasn't a sergeant from the third platoon of the Chornobyl police who visited us, but General Lynx. But what would sergeants do here in the backwoods, anyway? That day we reached the hamlet of Zolotniyiv. It does not exist in the wild—even the Kyiv Military and Cartography Factory has almost forgotten about it. I always search for the most remote spots and go straight there. After several years of wandering, illegal tourists feel no alienation whatsoever. It's only present in the distant border zones where illegal and official tourists never set foot. It's only present where oblivion is.

THE POLICE MIGHT be lying in wait, ready to ambush you, or they might not. I might notice the glow of a cigarette or the silhouette of their car. I might hear them speaking quietly in broken Ukrainian in the dark and change my route. Or I might not—then they'll catch me and grill me for a long time about where those five guys with whom I'd traveled to the barbed-wire fence by shuttle bus—contrary to my habits—are hiding. They'll steal my Swiss army knife for the third time and shove me into a patrol car and keep an eye on me, the stalker, until morning.

Actually, things can wind up being much more mundane, and the sergeants might nod off at 4:00 a.m., and I might slip by them. Like that day when I was running toward Prypyat when it was pouring, amid the blinding flashes of the thundering skies. If they are snoring big-time, I might pull out a sharpie and write a note on a piece of paper, something like, "Hi guys!" and stick it under their windshield wiper. Just hope they won't wake up right at that moment. The main thing is to have a philosophical attitude toward police reports and to remember that there's always a certain risk involved, although thoroughly planning your route through the impenetrable thicket reduces it to a minimum. That's police-reportism for you—a well-balanced mixture of stoicism and fatalism.

This wretched piece of land is not called Chornobyl Land, or Polissya, or the Zone of Alienation, or the Forbidden Zone. It's called Trespassingland. Honestly, I don't even remember how many times the men in blue have filed reports against me for trespassing. Sometimes, I just wander around till I'm blue in the face, staying in Prypyat for weeks, and then I turn myself in just for the fun of it. Or I

try to take the Chornobyl–Kyiv shuttle bus right at the checkpoint, by hiding behind the back row. I never get lucky, of course. A cop always steps inside the bus, checks everything thoroughly, and notices me. He asks me what kind of papers I have, I tell him I have a "report," he tells me to show it to him, and I tell him that I don't have it yet, but I will get it soon. He treats me to tea at the checkpoint, we exchange the latest gossip, and he sends me away with a warning—"Never, ever come back!" Of course, he doesn't believe it. Neither do I.

I've never said "Farewell" to the police officers—only "See you later." We'll see each other again, to be sure. Perhaps I'll get bored and I'll go to Chornobyl again. I'll go to the cafeteria—unshaven bum that I am—leave my backpack in the cloakroom, and sit down at the table for a hearty meal. I'll sit down, take a sip of borscht, look around, and these sullen men that chase after me in the woods will be sitting there chewing their boiled buckwheat as if nothing happened. Twenty of them. Half of them will have recognized me by that point. Well, hello.

I was looking for something unattainable among the hermits and self-settlers, on the wild boar paths and the smugglers' tracts, in the most secluded borderlands of the Zone, in the secret nooks of Polissya—at a benefit performance of all things absurd and the El Dorado of a handful of enthusiasts yearning to feel like Columbus, like pioneers of hamlets and forest districts. Such routes and perversions suit only those long bored with the Zone.

That's why I needed an inflatable boat and a ninety-pound backpack for my monthlong walk. I lugged that backpack in the scorching-hot sun; I lugged that boat with all my cats, fears, and worries. I had an urgent need to sail it on the cooling pond. The pond is an immense ocean next to reactor No. 4 of the Chornobyl NPP. Sailing across that pond on a boat is terribly arrogant and exotic. So I sailed.

The bees were buzzing, and the sun's rays flew into the rusted hangars of the floating factory through the bullet holes, the bees following them inside. I decided to go for a swim. I lay on the rocks and washed my clothes salty with sweat. Muddy and slimy, they solidified and changed color. Some official tourists were walking just a short distance away, and I tapped "The Call of Winter" on the skull of a huge fish and felt completely happy. Then I got fried in the sun. I smoked a couple of Camels and began inflating the boat. Seagulls were flying low over the water, and I was only thinking, "Fuck, why didn't I bring a proper camera?" A decent beginner-level reflex camera with a spare battery. I had such a great angle of the station. The old pipe was still there, and I could take a picture of it, capturing part of the cooling pond. The sun was going down, and I sailed closer to the Chornobyl NPP.

It was a farewell to the Old Zone, when the pipe—a symbol of all global disasters—was still in place. I thought that I might never get a chance to go up there and get an eyeful of it. I won't get a chance, I thought, and my nostalgia will

be forever tinged with this unfortunate sadness. But I did get a chance. The old pipe was dismantled before my eyes. Farewell, Zone. You are frozen time, and I will always see myself in you from now on.

ILLEGAL TOURISTS MAKE dead cities alive again. They breathe life into the empty shells of fragile houses and peeling apartment blocks. They make bright fires on apartment floors during dark nights, drink alcohol, play guitars, smoke cheap cigarettes, tell funny stories, laugh, and then snore loudly into the darkness. They make Prypyat alive, a place worth living for, a place you'll walk twenty-five miles to on a cold night, hiding in the dark from other people and cars.

Sometimes the Zone can be a truly dangerous place. Which trajectories and routes might lead you to trouble here? Which signs are good and allow you to bypass police ambushes and border checkpoints? What are the laws measuring how safe this or that route is and how many million impressions you can fit on a square inch of this poisoned soil? What are the chances that wolves won't tear me to pieces tomorrow? How much time do I have left to live, given the hundreds of liters of poisoned water I've drunk? I don't know.

One day, I'll get in serious trouble. The laws of trouble in the Zone are as simple as hell—it's hard to get yourself into a real mess, but once you do, you'll freeze to death deep in a swamp. But my backpack is packed. In half an hour, a plane with three Germans is supposed to land at Boryspil Airport. Although terrified by the whimsies of the Zone, they have still entrusted their well-fed bodies to me. I'll plop down on the front seat, slam the door shut, and a faint drizzle will crawl over the high-rises of the living, while I'll flee to the high-rises of the dead. The radio will be silent.

In three hours, I'll light a blue Camel and pull up to the barbed wire, and my German companions will sneak into the kingdom of emerald swamps. "Congratulations. You've just broken the law." That's all I will say. They'll ask me, "You come here so often. Aren't you afraid of radiation?" And I'll tell them, "No. It's only here that life won't slip by me, for I'm living it in the most exotic place on Earth."

When people ask me about my health, I really have no idea what to tell them. Yes, it's very harmful. It's a bad idea to drink water from poisonous lakes, utility tunnels, swamps, and ditches filled with scrap metal and the corpses of roe deer. But life often happens to us, and sometimes death does, too. Sometimes we're given good health, and sometimes we're harmed. And I firmly believe that, in two decades, I will meet those boys and girls who kept me company during my travels around the Zone in the chemotherapy room of a nice cancer clinic in Kyiv. And I know that we'll smile at each other. We'll smile at a life that challenges you and dictates where you should walk, how you should live, and what you should breathe. After all, we're the children of our time. Where else could we be?

It's time to go home. On this occasion, I walk unhurriedly around downtown Prypyat—my lawful domain. I

sit for a long while in the amusement park, stomp on broken glass in the spacious halls of the Energetik Palace of Culture, and then climb onto the roof of a building adorned with the Soviet coat of arms to give the City one last look.

I'm standing on the rooftop, looking at the abandoned building. At the skyline of the Zone's stone heart, at the skyline of Prypyat. The late-afternoon sky is warming up my tanned, smiling face, without burning it. Fleecy clouds stretch across the blue sky far away above the horizon. A cool breeze is blowing. Now I know for a fact—I will return again.

2012–2014

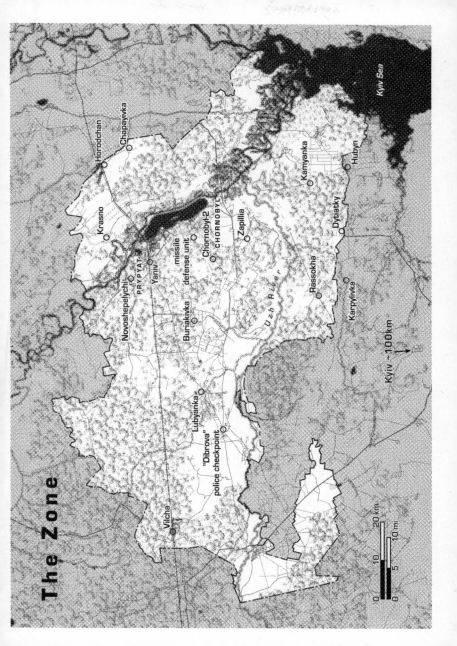

The Zone

Kyiv Sea

Horodchan
Chapayivka
Krasno
Kamyanka
Hubyn
Novoshepelychi
PRYPYAT
Yaniv
missile
defense unit
Chornobyl-2
CHORNOBYL
Zapillia
Dytiatky
Buriakivka
Rassokha
Karpylivka
U z h   R i v e r
Lubyanka
"Dibrova"
police checkpoint
Vilcha

Kyiv ~100km

0   10   20 km
0   5   10 mi

Acknowledgments

Eleonora Simonova, Mykola Kravchenko, Pierre Astier, Laure Pécher, Andrey Kurkov, Iryna Dmytrychyn for their faith in me and support of my career. To the whole team of my Ukrainian publishing house Nora-Druk and the whole team of my agency Astier-Pécher Literary & Film Agency. To Hanna Leliv and Reilly Costigan-Humes for the translation. To the whole team of Astra House for the preparation of the American edition of the book.

About the Author

Markiyan Kamysh is a Ukrainian writer who represents the Chornobyl underground in literature. Since 2010, he has been illegally exploring the Chornobyl Exclusion Zone. He is the son of a Chornobyl liquidator, nuclear physicist, and design engineer of the Institute for Nuclear Research in Kyiv who died in 2003. *Stalking the Atomic City* is his first book, which was translated in multiple languages and published to great acclaim. He lives in Kyiv, Ukraine. See more photos on his Instagram, @markiyankamysh.

About the Translators

Hanna Leliv is a freelance literary translator based in Lviv, Ukraine. In 2017 and 2018, she was a Fulbright fellow at the Iowa Translation Workshop. Among her translations into Ukrainian are texts by Kazuo Ishiguro, Ernest Hemingway, and Stephen Hawking. Her translations of contemporary Ukrainian literature into English have appeared in *Asymptote*, *Washington Square Review*, *The Adirondack Review*, *The Puritan*, and *Apofenie*.

Reilly Costigan-Humes translates contemporary Ukrainian and Russian literature. He is best known for his renderings of Serhiy Zhadan's novels into English.